Intergenerational Connections in Digital Families

Sakari Taipale

Intergenerational Connections in Digital Families

Springer

Sakari Taipale
University of Jyväskylä
Jyväskylä, Finland

ISBN 978-3-030-11946-1   ISBN 978-3-030-11947-8   (eBook)
https://doi.org/10.1007/978-3-030-11947-8

Library of Congress Control Number: 2018968371

This Springer imprint is published by the registered company Springer Nature Switzerland AG
The registered company address is: Gewerbestrasse 11, 6330 Cham, Switzerland

Acknowledgements

This book summarizes the main results of the research project Intergenerational Relations in Broadband Societies, funded by the Academy of Finland. The five-year project, running from 2013 to 2018, explored digital media and communication technology use among family generations and social generations (cohorts). Of particular interest for it were the ways in which new technologies might separate generations while perhaps also creating new connections between them, and whether there was any variation in this regard between countries of the European Union.

The bulk of this book is based on interview material collected in Finland, Italy and Slovenia in 2014 and 2015. Without the dedicated and diligent local co-researchers in charge of organizing the interviews, it would have been impossible for me to study families dispersed geographically across three countries and, quite concretely, interview them in three different languages. Observing and interviewing their own families and kin, my key informants produced notably rich research data that, in its subsequent English translation, I have had the great pleasure of revisiting over and over again for the benefit of my ongoing research.

In the course of the project, many colleagues supported my work, aiding in the data collection, sharing insights and giving valuable feedback. While all of their input and help was critical for the project's success, a few of them nonetheless deserve a special mention. Professor Leopoldina Fortunati and Dr. Manuela Farinosi have been my trusted partners for many years. For this project, they helped me with the recruitment of key informants in Italy and invited me to work on it as a Visiting Professor at the University of Udine in 2015. While in Udine, Associate Professor Vesna Dolničar and Assistant Professor Andraž Petrovčič stepped forward, volunteering to assist with data collection in Slovenia. Shortly afterwards, I had the privilege of being able to join their research team as a Senior Research Associate at the Centre for Social Informatics, University of Ljubljana. Dr. Farinosi, Dr. Dolničar and Dr. Petrovčič were also my co-authors on two articles that, with their kind permission, I have revised for inclusion in this book. I would, furthermore, like to mention my research visit to the Internet Interdisciplinary Institute (IN3) at the Open University of Catalonia, Spain, in 2017: it was during my stay there that the structure of this book was finalized and many of its chapters initiated. I would like to express my gratitude, in particular, to my two wonderful hostesses at the university, Dr. Mireia Fernández-Ardèvol and Dr. Andrea Rosales.

There are also many colleagues in my home department at the University of Jyväskylä, Finland, who need to be thanked. Some of them assisted me with data collection and analysis; others helped me to clear my tangled thoughts. Those deserving a special mention here include: Dr. Riitta Hänninen, Dr. Mia Tammelin, Dr. Tomi Oinas, Prof. Terhi-Anna Wilska and Mrs. Sanna Kuoppamäki, along with my former Research Assistants Emilia Leinonen and Armi Korhonen. Their assistance, support and advice have been invaluable. The project Digi50+: Mature Consumers, Customer Experience and Value Creation in Digital and Physical Environments, led by Prof. Wilska, provided a platform for exchanging thoughts and ideas about generations, ageing, life course and new technologies. As of January 2018, the university's Centre of Excellence in Research on Ageing and Care (CoE AgeCare), of which also my own research group, New Technologies, Ageing and Care, is part, has provided a new platform for these discussions and debates.

Finally, and most importantly, this book about digital families would perhaps never have seen the light of day without the support of my own family—all three generations of it, all online and connected.

Sakari Taipale

Contents

Part II   Roles, Responsibilities and Practices

Part III   Conclusions and Implications

Chapter 1
Introduction

Abstract In this chapter, the overall topic of the book and its rationale are introduced to the reader. The book develops an argument about the rise of digital families and examines how such families use different technologies to their diverse ends. Today, the lives of both their youngest and adult members are already highly 'connected' via portable and personal communication technologies. However, it is only now that the oldest family members are getting ready to engage in digital and online family interactions. The introduction ends with the presentation of the structure of the book.

Keywords Digital family · Extended family · Family solidarity · Generations · Information and communication technology · Linked lives · Technology adoption

Three Generations of Digital Technology Users

The digitalization of families started with their youngest members and young adults becoming early adopters of game consoles, personal computers and CD players in the late 1980s and 1990s. Later on, they did the same with MP3 players, mobile phones and many other tools and gadgets coming out on the market. North America and Western Europe were paving the way in this regard, although adoption rates and rhythms were notably varied even internally within these two regions. However, thanks to the steady advance of digitalization, the largest generational gaps in the uptake of most common personal communication and media technologies have continued to narrow everywhere in the Western world (e.g. for Finland, see Wilska & Kuoppamäki, 2018). Yet, at the same time, some new differentiating factors, bearing upon, for instance, the breadth and purpose of people's Internet use, have become more and more evident and apparently influential with time (Pearce & Rice, 2013; Taipale, 2016). As a result, scholarly attention has increasingly begun to be paid to the internal diversity of generational groups, when previously they were looked upon as basically homogenous by nature (Friemel, 2016; Hargittai & Dobransky, 2017).

Families of all sizes and shapes have become quickly saturated with digital devices, and today, the lives of both their youngest and adult members are already highly 'connected' via portable and personal communication technologies. However,

© Springer Nature Switzerland AG 2019
S. Taipale, *Intergenerational Connections in Digital Families*,
https://doi.org/10.1007/978-3-030-11947-8_1

it is only now that the oldest family members are getting ready to engage in digital and online family interactions. This, to be sure, is true in the first place of those among them who are male and have more education, higher income, a professional occupation and supportive family descendants (e.g. Friemel, 2016; Hargittai & Dobransky, 2017). Nevertheless, we can see a general trend that families consisting of up to three generations now become digitally increasingly connected.

This observation that families are quickly becoming digitalized is supported by statistical evidence from both Europe and the United States, even where the figures provided are typically household-based only (e.g. Kennedy, Smith, Wells, & Wellman 2008; Pew Research Center, 2017). The notion informing the data collection, that of a one-family household, fits poorly with the reality of numerous mixed and extended families made up of members regularly switching between households and belonging to many families at once. In the absence of more detailed family-level data, however, we must settle with conclusions and assumptions drawn from household and individual-level figures. In 2016, the share of households with Internet access in the European Union was already 85%, an increase of 30 percentage points from 2017 (Eurostat, 2017).[1] In 2016, only 14% of Europeans had never used the Internet, although there were pronounced country differences in this regard. The proportion of Internet non-users was still high in countries like Bulgaria (33%), Italy (25%) and Slovenia (22%), while it was significantly below the EU average (14%) in Nordic countries like Finland (4%).

The rates of adoption and use of basic digital technologies grow currently fastest in the oldest age group, especially among those towards the upper range of it. In Finland, for instance, the Internet use rates have continued to steadily rise among those aged 65 and older. As the figures for 2015 show, 27% of those aged 75–89 in the country had used the Internet in the preceding three-month period. Similarly, the proportion of Finns owning a mobile or a smartphone is already high across all age groups, except in the oldest age bracket; of those aged 75–89, no more than 5% possessed a smartphone in 2016 (Statistics Finland, 2017). Figures from the United States demonstrate a similar trend, showing that 65-year-olds now have caught up with their younger compatriots in their rate of broadband adoption (Anderson & Perrin, 2017). In 2016, 67% of those in the age group (65+) had used the Internet at least occasionally, compared to 90% of all adults younger than them. Any such differences between age groups tend, however, to be much more significant when it comes to ways of using digital technologies. While, in Europe in 2016, searching for information about services and goods and sending emails were the two most popular activities among Internet users aged 16–75, overall, the younger ones in this group were more frequently engaged in social networking (88% of those aged 16–24 did so, while the corresponding figure for the age group 55–74 years was 38%). Comparable differences were found in Europeans' Internet voice and video calling patterns as well as in their video watching and online content sharing habits (Eurostat, 2017).

In this book, I make the argument that, in the economically more developed societies, we are currently witnessing the emergence of digital families. In a dig-

---

[1] From here onwards, the term 'Europeans' is used to refer to the citizens of the European Union.

ital family, everyone from grandchildren to grandparents has at least some basic familiarity with information and communication technologies (ICT), knows at least some social media and has access to basic communication devices such as a mobile phone and the Internet, which one then uses to various degrees to stay in touch with other family and extended family members. In contrast to neighbouring concepts such as the 'networked family' (Kennedy et al., 2008) or the 'networked household' (Kennedy & Wellman, 2007), the concept of the digital family refers to the daily communication practices taking place within our extended and geographically distributed family relationships; that is, it covers not only child–parent but also parent–parent and child–grandparent relationships. While the concept will be subjected to a more thorough discussion in Chap. 2, it is worth noting already here that the emergence of digital families is an asynchronous and complicated process. It unfolds at different paces in different countries and regions, and the intensity of older people's participation in digital family life, using communication devices and social media applications, varies even greatly from place to place.

Digital families make possible a social structure in which personal communication technologies can be employed to serve not only individual aims but also the goals of sustaining family connections, caring relationships and family solidarity (Taipale, Petrovčič, & Dolničar, 2018). The debates on the productive versus counterproductive effects of personal communication technologies and media on family unity are long-standing, with ardent supporters on both sides (e.g. Shove, Pantzar, & Watson, 2012). Concerns have, for instance, been raised about the negative consequences of ICT and social media for family coherence. Some, like Turkle (2011), have argued that ICT substitutes for genuine interpersonal relationships, making us feel connected yet emotionally empty. Worries about diminishing family time have often focused on a trend towards privatized solo use of new personal technologies (e.g. Livingstone & Haddon, 2009). Others have suggested that technology may lead to forms of 'paranoid parenting' (Furedi, 2001) or 'helicopter parenting' (Clark, 2013) whereby parents constantly monitor their children's doings online, to avoid their becoming victimized by bullying, abuse and the like there. Kennedy and Wellman (2007), for their part, have proposed that personal communication technologies undermine the cohesive power of family solidarity that is based on normative expectations and reciprocity, and that, as a consequence, family solidarity is being replaced by loose ties between increasingly individually networked family members (see Chap. 8).

In other connections, to be sure, scholars have also demonstrated the positive consequences that digital technologies can have for family life. As one survey study from the United States, for instance, shows, many believe families to be more likely to stay together than grow apart thanks, precisely, to the impact of ICT (Kennedy et al. 2008). Other studies have demonstrated how new communication technologies can help family members to stay in contact with one another despite geographical and generational distances, especially upon the younger family members' leaving the nest (Epp & Price, 2008; Mesch, 2006; Tsai, Ho, & Tseng, 2011; Wilding, 2006). Interestingly in itself, in countries with a strong tradition of filial piety and family care provision, like China and Taiwan, the number of ageing parents living alone is today rising, making family members to more and more rely on online communication tools in their quest to organize family matters from afar (Tsai et al. 2011).

One the arguments made in this book is that the current-generation digital ICTs, in particular social media and various instant messaging applications, add to the range of technologies that can offer digital families new means for, and ways of, being together, providing care and maintaining emotional linkages beyond dyadic family relationships. The acts of community building that these technologies make possible can function as a balancing counterforce to the increasing prevalence of individual-centred networking via personal communication technologies, a phenomenon extensively studied and theorized in previous research (e.g. Rainie & Wellman, 2012; Wellman, 2001; Wellman et al., 2003). At the same time, however, new media and new communication technologies have also been shown to create tensions and frictions within families and between family generations (Taipale, Wilska, & Gilleard, 2018). The diversity and omnipresence of new ICTs and applications force family members to consider and discuss among themselves which digital tools should be utilized for their family communication, among which members, exactly, and to what extent, all from the point of view of how daily matters in the family can be organized and coordinated using new technology, from both near and afar. Without awareness of other family members' communicative preferences and digital skills, and without a shared agreement about which communication tools are suitable for just one's own particular family, disagreements and conflicts are inevitable.

To understand the digital connectedness of families, it is useful, though not enough in itself, to look at how 'generations' have been understood and conceptualized in classical sociology (see Taipale et al., 2018). Generation studies have sought to identify distinguishing factors that make one generation separate or different from another, in a process also known as generational 'othering' (e.g. Brown & Czerniewicz, 2010). In that context, generational differences are explained with reference to shared historical events experienced by members of a certain generation only. To study the intertwining of family members' lives, however, another kind of theoretical approach is needed. In response to that need, this book draws upon sociological theories of the life course and employs the concept of 'linked lives' (Cox & Paley, 1997; Elder, 1994; Elder & Kirkpatrick Johnson, 2003). Aided by these, it seeks to examine and describe how digital technologies can connect the lives of 'individually networked' family members, thereby shedding light on new kinds of interdependencies created among family members.

In the linked lives of families, digital media and communication technologies play a highly particular role. First of all, personal communication technologies, and instant messengers in especial, are increasingly being used to coordinate and synchronize the lives of individually networked family members (Ling, 2004; Ling & Yttri, 2002; Tammelin & Anttila, 2017). In geographically distributed extended families whose members' lives are characterized by high mobility, they offer an alternative to in-person communication needed to sustain intimate family connections. Second, these technologies can be used for 're-creating altered rituals and everyday interactions across geographically dispersed family members' (Epp & Price, 2008). In the absence of shared meals or, say, family gatherings around the television set, family members may share these moments with those physically not present via various technological means. Such small communicative acts may significantly contribute to the 'we' sense of families. Third, as Elder (1998) has reminded us, major events such as economic recessions that force families to change their lifestyles and spend money more sparingly, as well as important individual decisions needing to be made (such as about international work assignments or embarking on study abroad), affect family relationships. Also in such situations, digital families can resort to using new communication technologies to compensate for the lack of certain members' physical presence. Although new communication technologies cannot fully substitute for the physical vicinity of persons, it is good to also keep in mind that it is the very same technologies that enable individual and mobile lifestyles in the first place, and without them, the social costs of travelling and living apart would probably be much higher.

In developing its argument about the rise of digital families and examining how such families use different technologies towards their ends, this book draws upon data collected in empirical research, mainly qualitative but also quantitative. The principal research material consisted of written reports of observations and interviews conducted by key collaborators among their own extended families in 2014 and 2015, in three different countries—Finland, Italy and Slovenia. This material and the methods employed to analyse it are described in Appendix A.

The Structure of the Book

This book is organized into three parts, each consisting of two to four chapters. Chapter 7 and 8 are revised versions of a previously published book chapter (Taipale, Petrovčič, & Dolničar, 2018) and conference proceeding (Taipale & Farinosi, 2018), respectively. For Chap. 7, the analysis presented in the original article was extended so that it now covers also Italy, in addition to the original Finland and Slovenia. Chapter 8 reproduces the empirical analysis in the article on which it is based, but reframes it with new theoretical concepts and notions. Apart from some parts of Chap. 9 that have been published before (in Hänninen, Taipale, & Korhonen, 2018), the entire rest of the book consists of previously unpublished material.

Part I begins with a chapter dissecting the notion of the digital family, discussing its advantages and shortcomings to assess its usefulness vis-à-vis certain neighbouring concepts. In Chap. 3, the suggestion is made that the relevance and timeliness of this concept have to do with recent advances in personal communication technologies that, together with wider social changes in the developed world, facilitate older people's participation in a digital society. Next, in Chap. 4, the theoretical foundations on which the arguments in the book are built are developed. Particular attention is paid to the need here, based on the changes in the technological and social structure that we are witnessing, to have an approach that goes beyond strict generational dividing lines and is more sensitive to the ways in which individual lives are interconnected through the use of digital technologies.

Part II concentrates on empirical evidence of how the lives of individually networked family members are today in complex ways becoming increasingly interconnected with one another over the human life course. The analysis here focuses on family roles, responsibilities and practices that bind family members to, but also sometimes alienate them from, one another. Chapter 5 makes use of the insightful concept of 'warm expert', coined by Bakardjieva (2005) more than 10 years ago. Concrete examples of how the role of the warm expert is assigned in digital families, and how it is performed from near and afar utilizing different applications and mobile communication devices, are provided. In Chap. 6, the argument about the linked lives of the members of digital families is further substantiated, describing the way digital technologies have brought with them a need for a new type of housework: digital housekeeping. To sustain the functionality and reliability of the complex networks of personal technologies in the family (cf. Fortunati & Taipale, 2017), there is a new need to negotiate the fair distribution of the accompanying practices among the family members, taking into account each individual family member's digital skills, interests and other available resources such as money and time. Chapter 7 then takes a closer look at one specific mode of communication, WhatsApp messaging, used, especially in Finland and Italy, to facilitate communication in geographically distributed extended families. The benefits of the application are shown to be highest in countries where intergenerational linkages tend to otherwise be weak, loose or sporadic due to long geographical distances and early nest-leaving. Finally, in Chap. 8 an argument is made that the increasing use of new personal communication technologies and social media does not merely, or simply, erode family solidity: it can also be cohesion enhancing, strengthening linkages between family members and across generations. This cohesive force of digital technologies, however, stems in the first place from their association enabling and enhancing capability and the functional assistance that family members provide to those in need of help and support.

Part III then brings together and consolidates the empirical findings and theoretical constructs presented in Parts I and II. As is suggested in Chap. 9, in countries like Finland, in which families are geographically highly dispersed and rather loosely connected, digital technologies and social media, especially group messaging applications, can open up completely new avenues for family members to be more frequently in touch with one another, and thus for families to remain connected. The concept of re-familization, borrowed from the field of social policy research in which it refers to the growing responsibility of families to care for their loved ones on their own, is presented as an interpretative window through which to better understand this phenomenon. The book concludes with an outline, in Chap. 10, of a more balanced future approach that takes a more optimistic view of families and digitalization in our time. While digital media and communication technologies, undoubtedly, have some negative consequences for family relationships as well, personal communication technologies and social media are also something that individuals together can, and have already begun to, harness for their own, self-determined ends, to serve their common good as a family.

References

Anderson, M., & Perrin, A. (2017). *Tech adoption climbs among older adults*. Pew Research Center. Retrieved from http://www.pewinternet.org/2017/05/17/tech-adoption-climbs-among-older-adults/.

Bakardjieva, M. (2005). *Internet society: The internet in everyday life*. London: Sage.

Brown, C., & Czerniewicz, L. (2010). Debunking the 'digital native': Beyond digital apartheid, towards digital democracy. *Journal of Computer Assisted Learning, 26*(5), 357–369.

Clark, L. S. (2013). *The parent app: Understanding families in the digital age*. Oxford: Oxford University Press.

Cox, M. J., & Paley, B. (1997). Families as systems. *Annual Review of Psychology, 48,* 243–267.

Elder, G. H. (1994). Time, human aging and social change: Perspectives on the life course. *Social Psychology Quarterly, 57*(1), 4–15.

Elder, G. H. (1998). The life course as developmental theory. *Child Development, 69*(1), 1–12.

Elder, G. H., & Kirkpatrick Johnson, M. (2003). The life course and aging: Challenges, lessons, and new directions. In R. A. Settersten (Ed.), *Invitation to the life course: Toward new understandings of later life* (pp. 48–81). Amityville: Baywood Publishing.

Epp, A. M., & Price, L. L. (2008). Family identity: A framework of identity interplay in consumption practices. *Journal of Consumer Research, 35*(1), 50–70.

EuroStat. (2017). *Digital economy and society statistics—Households and individuals*. Retrieved from http://ec.europa.eu/eurostat/statistics-explained/index.php/Digital_economy_and_society_statistics_-_households_and_individuals.

Fortunati, L., & Taipale, S. (2017). Mobilities and the network of personal technologies: Refining the understanding of mobility structure. *Telematics and Informatics, 34*(2), 560–568.

Friemel, T. N. (2016). The digital divide has grown old: Determinants of a digital divide among seniors. *New Media & Society, 18*(2), 313–331.

Furedi, F. (2001). *Paranoid parenting: Abandon your anxieties and be a good parent*. London: Penguin.

Hänninen, R., Taipale, S., & Korhonen, A. (2018). Refamilisation in the broadband society—the effects of ICTs on family solidarity in Finland. *Journal of Family Studies*. Retrieved from doi.org/10.1080/13229400.2018.1515101.

Hargittai, E., & Dobransky, K. (2017). Old dogs, new clicks: Digital inequality in skills and uses among older adults. *Canadian Journal of Communication, 42*(2), 195–212.

Kennedy, T. L. M., Smith, A., Wells, A. T., & Wellman, B. (2008). *Networked families*. Washington, DC: Pew Internet & American Life Project.

Kennedy, T., & Wellman, B. (2007). The networked household. *Information, Communication & Society, 10*(5), 645–670.

Ling, R. (2004). *The mobile connection: The cell phone's impact on society*. San Francisco, CA: Morgan Kaufmann.

Ling, R. & Yttri, B. (2002). Hyper-coordination via mobile phones in Norway. In K. Katz & M. Aakhus, M. (Eds.), Perpetual contact: Mobile communication, private talk, public performance (pp. 139–169). Cambridge: Cambridge University Press.

Livingstone, S., & Haddon, L. (2009). *EU kids online: Final report*. London: LSE. Retrieved from http://www.lse.ac.uk/media@lse/research/EUKidsOnline/EU%20Kids%20I%20(2006-9)/EU%20Kids%20Online%20I%20Reports/EUKidsOnlineFinalReport.pdf.

Mesch, G. S. (2006). Family relations and the internet: Exploring a family boundaries approach. *The Journal of Family Communication, 6*(2), 119–138.

Pearce, K. E., & Rice, R. E. (2013). Digital divides from access to activities: Comparing mobile and personal computer Internet users. *Journal of Communication, 63*(4), 721–744.

Pew Research Center. (2017). *A third of Americans live in a household with three or more smartphones*. Retrieved from http://www.pewresearch.org/fact-tank/2017/05/25/a-third-of-americans-live-in-a-household-with-three-or-more-smartphones/.

Rainie, L., & Wellman, B. (2012). *Networked: The new social operating system*. Cambridge, MA: MIT Press.

Shove, E., Pantzar, M., & Watson, M. (2012). *The dynamics of social practice: Everyday life and how it changes*. London: Sage.

Statistics Finland. (2017). *Suomen virallinen tilasto (SVT): Väestön tieto- ja viestintätekniikan käyttö*. Retrieved from http://www.stat.fi/til/sutivi/index.html.

Taipale, S. (2016). Do the mobile-rich get richer? Internet use, travelling, and social differentiations in Finland. *New Media & Society, 18*(1), 44–61.

Taipale, S., & Farinosi, M. (2018). The big meaning of small messages: The use of WhatsApp in intergenerational family communication. In J. Zhou & G. Salvendy (Eds.), *Human aspects of IT for the aged population 2018* (pp. 532–546)., Lecture Notes in Computer Science Cham: Springer.

Taipale, S., Petrovčič, A., & Dolničar, V. (2018). Intergenerational solidarity and ICT usage: Empirical insights from Finnish and Slovenian families. In S. Taipale, T.-A. Wilska, & C. Gilleard (Eds.), *Digital technologies and generational identity: ICT usage across the life course* (pp. 68–86). London & New York, NY: Routledge.

Taipale, S., Wilska, T.-A., & Gilleard, C. (Eds.). (2018). *Digital technologies and generational identity: ICT usage across the life cours*. London & New York, NY: Routledge.

Tammelin, M., & Anttila, T. (2017). Mobile life of middle aged employees: Fragmented time and softer schedules. In S. Taipale, T.-A. Wilska, & C. Gilleard (Eds.), *Digital technologies and generational Identity: ICT usage across the life course* (pp. 55–68). London & New York, NY: Routledge.

Tsai, T. H., Ho, Y. L., & Tseng, K. (2011). An investigation into the social network between three generations in a household: bridging the interrogational gaps between the senior and the youth. In *Online Communities and Social Computing* (pp. 277–286). 4th International Conference, OCSC 2011. Dordrecht: Springer.

Turkle, S. (2011). *Alone together: Why we expect more from technology and less from each other*. New York, NY: Basic books.

Wellman, B. (2001). Physical place and cyberplace: The rise of personalized networking. *International Journal of Urban and Regional Research, 25*(2), 227–252.

Wellman, B., Quan-Haase, A., Boase, J., Chen, W., Hampton, K., Díaz, I. & Miyata, K. (2003). The social affordances of the Internet for networked individualism. *Journal of Computer-Mediated Communication, 8*(3).

Wilding, R. (2006). 'Virtual' intimacies? Families communicating across transnational contexts. *Global Networks, 6*(2), 125–142.

Wilska, T.-A. & Kuoppamäki, S. (2018). Necessities to all? The role of ICTs in the everyday life of the middle-aged and elderly between 1999 and 2014. In S. Taipale, T.-A. Wilska, & C. Gilleard (Eds.), *Digital technologies and generational identity: ICT usage across the life course* (pp. 149–166). London & New York, NY: Routledge.

Part I
Digital Families

Part I
Digital Families

Chapter 2
What Is a 'Digital Family'?

Abstract This chapter introduces the concept of the digital family. Digital families are one form of distributed extended families, consisting of related individuals living in one or more households who utilize at least basic level information and communication technologies and social media applications to stay connected and maintain a sense of unity. The strengths and limitations of the notion are discussed, assessing its usefulness vis-à-vis neighbouring concepts. The chapter ends with the discussion of the perception of family in the three countries studied, Finland, Italy and Slovenia, and of the differences found between them.

Keywords Digital family · Distributed family · Extended family · Information and communication technology · Network family · Social media

When reviewing research on new media and communication technologies, it quickly becomes evident how much scholarly attention has been given to dyadic communication practices in one-house families, especially among young people. In comparison, geographically distributed multi-household families, often consisting of several generations, have been left on the sidelines (e.g. Shove, Pantzar, & Watson, 2012). Family ties, however, tend normally to extend beyond the walls of a single household (Borell, 2003). Thanks to new digital media and communication technology, these distributed families can today nevertheless remain connected and feel a sense of togetherness, even when their members are not physically close to one another.

A large body of literature has explored media and technology use among children and young people, with especially the changing models of parenting in this regard attracting much interest (e.g. Lamish, 2013; Livingstone, 2002, 2009; Singer & Singer, 2012). The reasons behind this rather single-minded research focus are fairly obvious and quite understandable: children and young people are particularly vulnerable in the online environment, due to their cognitive and psychological immaturity and their relative lack of ability and experience. More recently, studies have, however, also begun to pay attention to middle-aged and older adults as users and consumers of personal communication technology and social media (e.g. Comunello, Fernández Ardèvol, Mulargia, & Belotti, 2017; Friemel, 2016; Ivan & Fernández-Ardèvol, 2017; Kuoppamäki, Taipale & Wilska, 2017; Kuoppamäki, Wilska & Taipale, 2017; Tsai, Ho, & Tseng, 2011). In many of these studies, which present older adults as a

© Springer Nature Switzerland AG 2019
S. Taipale, *Intergenerational Connections in Digital Families*,
https://doi.org/10.1007/978-3-030-11947-8_2

heterogeneous group of technology users, individuals' differing personal needs for, and ways of using, new media and communication technologies have been highlighted. Older adults have, among other things, been found to have become better equipped, more skilled and more interested in putting technological advancements to use for their ends, following their recognition of how new technology may facilitate their daily chores and help sustain their social relationships after retirement transition, with the old age approaching (Taipale, Wilska, & Gilleard, 2018).

Yet, despite this focus on individual persons in the families, families as such have not completely fallen under researchers' radar. There have been, for instance, attempts to address the effect of digital technologies on the lives of families, perhaps the most notable in Rainie and Wellman's *Networked: The New Social Operating System* (2012). In the book, the authors describe *networked families* as a social structure that provides families 'with a great deal of individual discretion, abundant opportunities for communication, and flexibility in their togetherness' (Rainie & Wellman, 2012, p. 147). While, we are reminded, the networking of families indeed began already prior to the ICT revolution, it was nevertheless not until the arrival of personal ICTs that wired (landline) phone calls and visits to people's homes, made to contact the entire household as a collective unit, were transformed into person-to-person communication events that subsequently replaced them. As Kennedy and Wellman (2007) have pointed out, however, also households have become more networked. New communication technologies have enabled family members to live their individual lives and go in different directions while still remaining connected, often even more than before, via mobile communication tools.

The *networked* or *connected home* is a concept closely related to the idea of the networked family. Venkatesh, Kruse, and Shih (2003) have defined it as a living space with multiple centres of activity (entertainment, work, communication, learning, etc.), which can be structurally divided into social, physical and technological spaces (e.g. Little, Sillence, & Briggs, 2009). Initially, the aim of the connected-home approach was to show the pitfalls of the then-current research agenda on smart homes, which stressed the multiplicity of the ways in which the domestic space was connected beyond the four walls of the home (Harper, 2011). With the concept of the connected home, attention was drawn to the power relations among family members and the power geometry within the domestic space. What remains unaddressed in these studies, however, are any technology-mediated and technology-related connections between family members who live in separate household or switch between two or more households while still perceiving themselves as members of a single-family unit.

Another major contribution to technology and family studies is made by Neustaedter, Harrison, and Sellen (2013), whose edited volume *Connecting Families: The Impact of New Communication Technologies on Domestic Life* explores the new ways family members connect with one another, and not only within the same household but also across distances and borders. The scope of the investigation extends beyond pure analysis of family networks, demonstrating the importance of the sense of connection for the identity of being part of the same family. Acknowledging that one often belongs to multiple families at once, and making a clearer distinction between 'family' and 'household' than what we can find in, for instance, Rainie and Wellman (2012), the book spotlights the role of grandparents (cf. Moffat, David, & Baecker, 2013) who, living elsewhere in other households, nevertheless play an important part in the life of the (extended) family. In families made up of several households and consisting of more than two generations (parents and their children), technology-mediated communication is shown to often serve families' need of staying connected, with the informational content of the communication being of no more than secondary importance.

The concept of the networked family thus offers a good starting point for understanding the digitalization of family relationships. Nevertheless, its historical rootedness in one-to-one communication technologies may no longer be suitable in today's world characterized by a wide array of communicative practices and patterns that extended families, sometimes living in multiple households, make use of to sustain family connections over temporal and spatial distances. The terms *distributed family* (Christensen, 2009) and *multi-household families* (Borell, 2003) describe such families perhaps more accurately. A distributed family is a variant of modified extended families, consisting of related family units born out of children moving out (Litwak, 1960). Yet, despite living far from one another and constituting households of their own, the members of a distributed family can continue to engage in, and develop, common family activities, and on a regular basis at that, by either visiting one another or using communication technologies for their purposes (see Browne, 2005)

Studies exploring the role of digital communication technologies beyond both dyadic family relationships and one-household families are, to be sure, not many. Judge, Neustaedter, and Harrison (2013), however, have carried out work on domestic media spaces specifically created for intra-family interaction, which they call the *Family Window* and the *Family Portals*, in an effort to understand how these can foster communication in modified extended families. In addition, they have provided a useful overview of other technologies developed for messaging between two or more households. Among these are platforms such as *commuteBoard* (see Hindus, Mainwaring, Leduc, Hagström, & Bayley, 2001), *messageProbe* (Hutchinson et al., 2003) and *Wayve* (Lindley, Harper, & Sellen, 2010). In Taiwan, Tsai, Ho, and Tseng (2011) have examined communication within three-generation households, finding that, in addition to face-to-face interaction, communication via telephone, email, instant messaging and social network sites contributed to family socialization, and that also older family members began to gradually use the Internet more, to establish and maintain contact with their children and grandchildren. Similar results have also been obtained in Estonia, where Siibak and Tamme (2013) have studied web-based communication tool use in three-generation families. What they found was that digital tools considerably facilitated intergenerational communication, over distances but also within the same household, and helped to revive intergenerational communication that had attenuated. In the section that follows, the concept of digital families is more systematically introduced and defined, to help us better understand how digital media and communication technologies are interwoven with the daily life of families of three or more generations that live in either one or several interconnected households.

What Makes a Digital Family?

Digital family, as defined for the purposes of this book, is one form of distributed extended family, consisting of related individuals living in one or more households who utilize at least basic information and communication technologies and social media applications to stay connected and maintain a sense of unity despite no more than occasional in-person encounters between them. Families of this type are, in fact, only now developing and becoming visible, after older family members, grandparents in particular, have begun to adopt and make use of a larger variety of digital technologies for family communication. Although person-to-person communication via mobile phone calls and text messaging, which formed the foundation of networked individualism (see Rainie & Wellman, 2012), has already established its position in families, today's mobile and social media applications offer novel avenues for group-based family communication to develop.

Digital families represent the outcome of family members' collective actions, including both deliberate and non-deliberate use of personal and mobile communication technologies to nurture family relationships. Besides direct technology-mediated communication, also digital appliances and software applications tie family members together, thus requiring intergenerational and intra-generational collaboration in the maintenance of the digital home. The new forms of intra-family collaboration range from actions taken to coordinate new digital hardware purchases and installation to those around configuring, updating and recycling the equipment.

In this connection, it is important to note that what allows a family to become, specifically, a digital family is its flexible social structure. Given the fluidity of contemporary human relationships and our increasingly non-standard personal biographies, also our family compositions tend to become more changeable over time. Who the persons making up our family are can thus change, even several times in the course of an individual life (cf. Finch, 2007; Venkatesh, Dunkle, & Wortman, 2011). This can happen with the ageing of the family-forming group, with family members getting married and separating, and with new members being introduced to the family group through births, remarriages or new partners. Owing to this structural instability, I argue below, any digital family should be seen as a changeable configuration that keeps being shaped and reshaped by both family members and non-members.

Membership and Composition

Family membership can be defined in different terms, based on, for example, legal, biological/genetic or affective membership, or any combination thereof. In Western societies, family membership has traditionally been viewed through the lens of law, as configurations of legal relationships translating into obligations divided between parents, children and the state. These obligations range from those relating to custody and the responsibility to provide maintenance to minors and secure their physical, mental and financial well-being, to those entailed by the biological/genetic and marital relationship in the form of, for instance, inheritance rights after a family member's death (Olivier & Wallace, 2009). As Olivier and Wallace (2009, p. 205) have shown, human–computer interaction research typically takes these formal and statutory frameworks as its starting point, making families appear as pure 'functional units that work, eat, relax and partake in leisure activities, more or less together'. Conceiving the matter this way is clearly not very productive for those wishing to understand how the practices of technology use shape digital families.

Family membership may, however, also be regarded as based on affective bonds between family members. Unlike that derived from statutory obligations, affect-based family membership comes about as a result of shared everyday life and mutual inter-actions. A significant part of that life and those interactions is, however, today medi-ated by personal media and communication technologies. As research has shown, personal communication tools, especially mobile phones, also serve as an important reservoir of personal and family memories (Oksman & Turtiainen, 2004; Vincent, 2006; Vincent & Fortunati, 2009). Sharing emotions forms a crucial part of affective relationships, and of our personal communication tool use mode (Lasen, 2004), as it enables the creation and sustenance of loose bonds that can be easily untied when needed or wanted. Some of the affective bonds may be widely shared and electroni-cally mediated within the extended family, while others might connect only a small group of people to one another, such as one parent and her or his biological child or children in a mixed family.

Like all other families, digital families are diverse in terms of their size and their gender and ethnic composition.[1] Key family members may be suddenly lost to unexpected death, or changes in the family composition may occur more predictably, such as when children grow up and move away from parents to start a family of their own, or couples decide to apply for a final divorce following a mandatory reconsideration period. In such and other cases, the addition of new, and the loss of current, family members prompts the remaining family to reconsider and adjust their ways of using technology for communication. In the process, those involved in it must consider 'again and again whom to include as members of their family' (Epp & Price, 2008, p. 52).

Very often in statistical research and official statistics, any changes in household size are taken to be indicators of transformations in family structure, even though household data cannot fully reflect either the diversity of families or differences in individual perceptions about who 'belongs to my family'. At the same time, however, the household is not entirely irrelevant for the way the family composition is per-ceived, either. The immediate household shapes family practices and has functional ramifications for the daily life of the digital family. For instance, the presence of chil-dren in the household promotes older family members' use of digital technologies (Kennedy, Smith, Wells, & Wellman, 2008; Lin, Tang, & Kuo, 2012; Luijkx, Peek, &

---

[1] Any more detailed discussion of different family forms falls outside the scope of this book; see, instead, e.g. Ciabattari (2016).

Wouters, 2015; Mori & Harada, 2010). In 2011, Hamill showed computer adoption in the United States to be influenced primarily, not by money, but by the presence or absence of children in the household (Hamill, 2011). Elsewhere, in Latin America, Cáceres and Chaparro (2017) have found that while the presence of young people in the households promoted older adults' Internet adoption to begin with, the presence in them of their spouses or partners increased the time these adults actually spent online. Correspondingly, when family members do not share the same household, older family members appear to learn and adopt digital technologies far more slowly or reluctantly (Taipale, Petrovčič, & Dolničar, 2018).

Although extended families likely represent the most common family type in many Western countries today, we still lack relevant family statistics to enable a full picture of their diversity to emerge (cf. Browne, 2005, p. 92). Official European data in the area is limited to first-degree family relationships, counting only the relationships between parents (of either sex) and their (blood, step or adopted) children (Eurostat, 2015). At the same time, the household statistics (Eurostat, 2015, 2017; Oláh, 2015) provide a rather unambiguous view of the changes either taking or having already taken place in the family composition. In Europe, the overall trend has been towards smaller households, owing to the decrease in the number of extended families living in the same household and the growing share of people, both young and the elderly, who live independently, along with declining fertility rates and increasing divorce rates. In 2013, single-person households accounted for about one third (32%) of all private households in the EU28 group of countries. In the same region, the share of households consisting of one or two persons rose from 59 to 63% between 2005 and 2013. What one also should note here, however, is that regional differences remain quite notable within the EU. Single-person household are more numerous in Northern Europe (making up, e.g. 41% of all households in Finland in 2016) than in Southern and Eastern Europe (30% in Slovenia in 2016, 32% in Italy in 2015; see Eurostat, 2015, 2017).

To summarize, we can thus make the observation that families in Western societies have become less stable in their structure and more diverse in their composition than before. This circumstance forces digital families to constantly take up and think over the issue of which digital technologies and applications they should use for family communication. Adjustments to the established modes and manners of communication may also be inevitable when the composition of the family changes.

Doing Digital Family

Based on what we just learnt above, it makes sense to conceive of a digital family as more of a process of 'doing' than just 'having' or 'being' something (Morgan, 2011). Digital families are transformed and reshaped as old and new family members move in and out of them, and as these members start and stop making use of certain communication technologies together. Digital communication thus both constructs and reflects all the different configurations of family relationships that we see today.

Accordingly, Lim (2016) has described intra-family communication as a process of 'doing' family, whereby a relational culture is constantly created, sustained, recreated and redefined both vis-à-vis one another and through the mediation of technology. From this perspective, a digital family is an endless work in progress that will never be completely finalized or fixed.

Digital media and communication technologies enable doing *family* in a context where family members' daily schedules and routines are very different and hence difficult to synchronize. Doing family via new communication technologies, however, involves not just a joint effort by the children and their parents: it engages all family generations. Also, grandparents partake in it and that regardless of whether they share the same household or not. Doing family, moreover, can also take the form of a *skipped-generation communication* whereby children and their grandparents are directly in contact with each other, without the parents' involvement. In fact, grandparents play a crucial role in the 'doing' of extended families. As research has shown (Tsai et al., 2011), senior family members often act as family historians, advisors, nurturers and surrogate parents within the overall framework of the extended family. Grandparents represent a kind of 'reserve army', supporting both parents and their children when the family faces a crisis, such as in the form of a severe illness or unemployment, or when parents get divorced. Grandparents also help younger family members to see and place themselves in a long historical continuum of technology use. This they do, for instance, by sharing memories about the domestication of first home electronic devices and how these were used together and shared in their own family. Very commonly, the youngest family members today have little or no knowledge of what family life was like before, when there were no personal communication technologies such as smartphones and tablet computers. Understanding the generational differences in the experiential component of these technologies' use is, however, critical in bringing family generations closer to one another. The sense of belonging in a family is created through communication, and this sense endures principally only when family members work together to sustain it jointly.

Considering the formative role of senior family members and (other) family members not sharing the same household with rest of the family, it seems obvious that the identity of a distributed extended family can only come about and be established as a result of a collective effort in a shared process. In the digital family, technology-mediated interactions and technology-related family discussions lay foundations for, and shape, the family's 'we' sense. As Epp and Price (2008, pp. 50–51) have stressed, '[a]s families construct identity, they face competing interests and demands, increasingly elective and fluid interpersonal relationships, and blended family forms that depart from prevailing ideals'. In that situation, family identity then emerges as a combination of individual experiences, family relationships, and a collective 'we' sense. It makes possible for family members to reflect on who 'we' are as a family and in what respects that 'we' differs from the 'we' of other families (cf. Bennet, Wolin & McAvity, 1988). This is so also in terms of the family's technology use.

Digital families, accordingly, represent a diverse set of distributed extended families, made up of two or more generations that use new media and communication tools as well as social media applications to sustain and even revive family ties. Starting

out from this definition, my aim in this book is to promote thinking that deviates from
that represented by the individual networking and one-household approaches. The
focus in that effort is on intergenerational communication practices as they appear
in distributed extended families, in which children, parents and grandparent are all,
even if differently, engaged in the use of ICTs and social media applications.

The Perception of Family in Three Countries

The families in the three countries studied for this work, Finland, Italy and Slovenia,
differed considerably in terms of their size, shape and technology use patterns. In
general, what is regarded as a 'family' is both an individual and a cultural question,
and the answer to it can also change over time. As a result, the concept of the
family varies even greatly, referring to many things from a mini-group of two persons
(adult–adult or adult–child) to large extended and mixed families involving multiple
generations and a number of distant relatives.

In Europe, the main difference in the family concept is typically taken to be that
between 'Northern' and 'Southern' societies (Jokinen, 2014). It is, for instance, considered characteristic of the Nordic countries that the link between marriage and
family formation has considerably weakened in them. In contrast, a more traditional
family model based on marriage appears still relatively strong in Southern European
contexts. Also internally within some countries, such 'regional' differences in the
perception and meaning of family can be seen to be in evidence. As Piumatti and
collaborators (2016), for instance, have noted, the better employment and educational opportunities that, say, people in Northern Italy can enjoy favour generational
transitions, leading to looser psychological and economic family bonds and smaller
family units, compared to the country's south where large family units are still what
provide individuals with many of the affective bonds and social safety-net functions
that they need.

The key informants of this study supplied, among other things, also their own
family definition, describing who belonged to their particular family. The definitions
they gave typically reflected the general notions prevailing in their respective countries. Accordingly, as most of the Italian key informants were from the country's
North, another one of them, Emilio (aged 30), who was born and raised in its South
instead, made a point of noting how his view of 'family' differed from that of his
colleagues:

> Coming from southern Italy, my own personal experience is that of the 'classic' extended
> family. The way I see it, to my family belong also numerous uncles, aunts, and cousins of
> different ages. The family bond is stronger on my maternal side, though, as we've spent a
> lot more time together in the last couple of years.

In general, the Italian key informants counted as part of their families not only
their parents and siblings but also their grandparents and cousins. As appears from the
quote above, sometimes also aunts and uncles could be included. Another informant

presenting such a broad notion of family was Bruno (aged 30), who reported that 'My family consists of my parents, my brothers and sisters, their companions and their children, my grandparents—although they have deceased—as well as my uncles and cousins'. The Italian key informants' close relationship with a wide range of their relatives was also reflected in their selection of interviewees for this study (see Appendix). Nineteen out of the 21 informants in Italy included also family members other than their parents, siblings and grandparents in their fieldwork. In contrast, only four out of the 22 key informants in Finland conducted 'family member' interviews with their cousins and/or aunts, with everyone else restricting them to their parents and siblings only, albeit including also stepparents, stepsiblings and adoptive siblings in these categories. Also, co-habiting partners and, sometimes, parents-in-law and siblings-in-law could be defined as family members.

Compared to their Italian counterparts, the Finnish key informants thus defined their family more narrowly, likely reflecting the distinction drawn in the Finnish language between the kinship terms *perhe* and *suku*. Of the two, the former covers only the closest family community, while the latter refers to all blood relatives. A typical Finnish definition of a family (*perhe*) was provided by Jenny (aged 25): 'As I define it, it's my father, my mother, and my sister who make up my family, and my boyfriend, too. To me, my relatives or my partner's family don't fall under the category of "my family"'. There were, however, a couple of other key informants in Finland who included their grandparents in their concept of family. One of them was Emma (aged 24), who, to be sure, also herself noted that hers was an unusually broad family definition for the Finnish cultural context. Another was Marika (aged 29), who stated that:

> My notion of who belongs to my family is quite broad. Of course, there is the core family that includes my mother, my father, and my brother, but I also think my grandparents are part of the family. My boyfriend has also become part of my family in the course of our long dating period.

Also, Benjamin (aged 29) in Finland regarded his grandparents as part of 'the family', even if he drew a small distinction between them and his other, core family members: 'Both of my grandparents and my sister-in-law, too, are people who, to me, are almost comparable to family members'. Interestingly, the Finnish key informants who lived in blended families did not present any broader definitions of their families along these lines, apart from including stepparents, stepsisters and stepbrothers in them (e.g. Ella, aged 24; Laura, aged 29).

Similarly in keeping with this narrower family conception, the Finnish key informants, furthermore, tended to divide their families into subunits, or inner and outer family circles. Julia (aged 21), for instance, did so when stating as follows:

> My family consists of my boyfriend and a cat that we live together with. In addition, also my mother, my sister, and my brother belong to my family, all somehow a bit differently. My other family ties are pretty loose, and if someone were to ask me about them, I wouldn't, for example, define my father or my grandma as part of my family.

In the same vein, Rita (aged 34) described that 'I think I have two families. My own family is made up of my husband and my five-year-old daughter. But to my extended family belongs also my childhood family—my mother, my father, and my sister, both of my grandmothers, and my sister-in-law'. This inclination to view one's family as consisting of separate units is probably, at least in part, attributable to the country's public welfare system, which, providing economic stability to support the basic needs of citizens from early on, makes it possible for children to move out of the parental home relatively early in their lives. Setting up one's own home at a very early age has become a strong cultural marker of independence acquired from one's natal family.

The Finnish research material also contained some examples of blended and atypical families. Laura (aged 29), for example, lived in a family that had become blended already twice. In addition to her, her family was made up of her mother and two stepfathers, a stepbrother and a younger stepsister. Another example of the diversity of families was Teresa (aged 24), whose family comprised 17 members in all, including not only the key informant herself along with her siblings, parents and three half-siblings from her parents' previous marriages, but also the spouses of her two siblings, her parents' five foster children, her stepbrother's children and a grandmother.

Just as in the Finnish cases, also the key respondents in Slovenia tended to distinguish between their immediate family and their extended family. Most typically, the immediate family was described as consisting of one's parents, siblings and grandparents, but sometimes also of one's aunts, uncles, and cousins (e.g. the families of Veronika, aged 27, Sandra, aged 25, and Alexander, aged 24). To their extended families some key informants, such as Marija (aged 25), included 'certain other members of my wider family, like my mother's cousins and the children and grandchildren of my mother's uncle'. The Slovenian key informant Jakob (aged 26) drew the distinction between his immediate and extended family rather straightforwardly:

> The way I look at it, I can say that my family is divided into immediate and extended family. The close family members all belong to my immediate family—my grandmother, my grandfather, my mother, my father, my brother, and my sister—as do also the partners and children of my brother or sister—my nephews and nieces—and my own partner—later, of course, our children as well. My extended family consists of the relatives of my parents—their aunts, uncles, cousins—and my partner's relatives.

Some Slovenian key informants also referred to an inner family circle of theirs when speaking of their most important family relations. Anton (aged 29), for instance, spoke of how 'I include my father and mother in the inner circle of the family. My immediate family includes the uncles, aunts, and cousins with whom I am in regular contact'. The key informant Tina (aged 25), for her part, had already established her own family, after marrying and starting to live together with her husband. Nevertheless, unlike her Finnish counterparts, she still regarded her immediate family as also including her parents (mother and father) and her sister, as well as her grandfather, grandmother, uncles, aunts and the latter's children (cousins).

The most significant and distinctive feature of the Slovenian families in this study was that the criterion used for defining them as such was still to a notable extent based on the notion of family members' physical vicinity to one another. There was resonance here with the tradition of multi-generation households and high preference for homeownership, often visible in the tendency to build large houses specifically for many family generations to live in, or to cluster multiple single-family houses on one and the same property owned by the family (Cirman, 2006; Mandić, 2001). The idea of the essentiality of the physical closeness of others that this tendency reflects was conspicuously present in the reports by Slovenian key informants. For them, their 'family' was made up of those whom 'we see often, meaning we go visit them or they come and visit us' (Marija, aged 25) or 'who live in my immediate vicinity and who we socialize with frequently' (Tina, aged 25; a similar definition was put forth also by Sebastjan, aged 26, and others). Correspondingly, long distances were cited as a reason for not regarding some people, even one's grandparents, as family members. As Petra (aged 25) put it, 'I don't consider my grandparents as part of my immediate family, because they live so far away and we therefore don't have such close contact with them'.

As noted already at the outset, the key informants' understandings of what constituted 'family' for them varied notably from country to country. The Finnish key informants' markedly narrow concept of it, combined with the ubiquity of blended families, is well in line with the popular notion that Northern countries have some of the least marriage-centred family cultures in Europe. Slovenia was in this regard the opposite to Finland, yielding a picture of large, multigenerational families for which the idea of spatial proximity of family members was important. Nevertheless, the Slovenian key informants' inclination to classify their families into the immediate and extended family members—the family's inner and outer layer—and the presence of many divorced parents among the interviewees indicate there to have been an already visible broader trend towards the pluralization of family forms also in Slovenia. The Italian families, as noted above, were considerably larger than their Finnish counterparts, typically involving also aunts, uncles and cousins, but, as in the Finnish cases, they were much more geographically distributed than the Slovenian ones.

References

Bennett, L., Wolin, S., & McAvity, K. (1988). Family identity, ritual, and myth: A cultural perspective on life cycle transitions. In C. Falicov (Ed.), *Family transitions* (pp. 211–234). New York, NY: Guilford Press.

Borell, K. (2003). Family and household: Family research and multi-household families. *International Review of Sociology, 13*(3), 467–480.

Browne, K. (2005). *An Introduction to Sociology* (3rd ed.). Cambridge: Polity Press.

Cáceres, R. B., & Chaparro, A. C. (2017). Age for learning, age for teaching: The role of intergenerational, intra-household learning in internet use by older adults in Latin America. *Information, Communication and Society, 22*(2), 250–266.

Christensen, T. H. (2009). 'Connected presence' in distributed family life. *New Media & Society, 11*(3), 433–451.

Ciabattari, T. (2016). *Sociology of families: Change, continuity, and diversity*. Los Angeles: Sage.

Cirman, A. (2006). Housing tenure preferences in the post-privatisation period: The case of Slovenia. *Housing studies, 21*(1), 113–134.

Comunello, F., Fernández Ardèvol, M., Mulargia, S., & Belotti, F. (2017). Women, youth and everything else: Age-based and gendered stereotypes in relation to digital technology among elderly Italian mobile phone users. *Media, Culture and Society, 39*(6), 798–815.

Epp, A. M., & Price, L. L. (2008). Family identity: A framework of identity interplay in consumption practices. *Journal of Consumer Research, 35*(1), 50–70.

Eurostat. (2015). *People in the EU—Who are we and how do we live? Statistical book.* Retrieved from http://ec.europa.eu/eurostat/statistics-explained/index.php/People_in_the_EU_-_who_are_we_and_how_do_we_live%3F.

Eurostat. (2017). *Digital economy and society statistics—Households and individuals.* Retrieved from http://ec.europa.eu/eurostat/statistics-explained/index.php/Digital_economy_and_society_statistics_-_households_and_individuals.

Finch, J. (2007). Displaying families. *Sociology, 41*(1), 65–81.

Friemel, T. N. (2016). The digital divide has grown old: Determinants of a digital divide among seniors. *New Media & Society, 18*(2), 313–331.

Hamill, L. (2011). Changing times: Home life and domestic habit. In R. Harper (Ed.), *The connected home: The future of domestic life* (pp. 29–57). Dordrecht: Springer.

Harper, R. (2011). From smart home to connected home. In R. Harper (Ed.), *The connected home: The future of domestic life* (pp. 3–18). Dordrecht: Springer.

Hindus, D., Mainwaring, S., Leduc, N., Hagström, A., Bayley, O. (2001). Casablanca: Designing social communication devices for the home. In *ACM SIGCHI Conference on Human Factors in Computing Systems 2001* (pp. 325–332). New York, NY: ACM.

Hutchinson, H., Mackay, W., Westerlund, B., Bederson, B., Druin, A., Plaisant, C., …, Eiderback, B. (2003). Technology probes: Inspiring design for and with families. In *Proceedings of the SIGCHI Conference on Human Factors in Computing Systems* (pp. 17–24). New York: ACM.

Ivan, L., & Fernández-Ardèvol, M. (2017). Older people and the use of ICTs to communicate with children and grandchildren. *Transnational Social Review, 7*(1), 41–55.

Jokinen, K. (2014). Families and family policies in finland: A future scenario. *Gender Studies and Policy Review, 7*, 36–54.

Judge, T. K., Neustaedter, C., & Harrison, S. (2013). Inter-family messaging with domestic media spaces. In C. Neustaedter, T. Harrison, & A. Sellen (Eds.), *Connecting families: The impact of new communication technologies on domestic life* (pp. 141–157). Dordrecht: Springer.

Kennedy, T., L., Smith, A., Wells, A. T., & Wellman, B. (2008). *Networked families.* Washington: Pew Internet & American Life Project. Retrieved from http://www.faithformationlearningexchange.net/uploads/5/2/4/6/5246709/networked_family_-_pew.pdf.

Kennedy, T. L., & Wellman, B. (2007). The networked household. *Information, Communication & Society, 10*(5), 645–670.

Kuoppamäki, S.-M., Taipale, S., & Wilska, T.-A. (2017). The use of mobile technology for online shopping and entertainment among older adults in Finland. *Telematics and Informatics, 34*(4), 110–117.

Kuoppamäki, S.-M., Wilska, T.-A., & Taipale, S. (2017). Ageing and consumption in Finland: The effect of age and life course stage on ecological, economical and self-indulgent consumption among late middle-agers and young adults between 1999 and 2014. *International Journal of Consumer Studies, 41*(4), 457–464.

Lamish, D. (Ed.). (2013). *The Routledge international handbook of children, adolescents and media.* London & New York, NY: Routledge.

Lasen, A. (2004). *Affective technologies—Emotions and mobile phones.* Receiver, Vodaphone, 11. Retrieved from http://www.academia.edu/472410/Affective_Technologies._Emotions_and_Mobile_Phones.

Lim, S. S. (Ed.). (2016). *Mobile communication and the family.* Dordrecht: Springer.

Lin, C. I. C., Tang, W.-H., & Kuo, F.-Y. (2012). "Mommy want to learn the computer": How middle-aged and elderly women in Taiwan learn ICT through social support. *Adult Education Quarterly, 62*(1), 73–90.

Lindley, S., Harper, R., & Sellen, A. (2010). Designing a technological playground: a field study of the emergence of play in household messaging. In *Conference on Human Factors in computing Systems 2010* (pp. 2351–2360). New York, NY: ACM.

Little, L., Sillence, E., & Briggs, P. (2009). Ubiquitous systems and the family: Thoughts about the networked home. In *Proceedings of the 5th Symposium on Usable Privacy and Security* (pp. 6–9). New York, NY: ACM.

Litwak, E. (1960). Occupational mobility and extended family cohesion. *American Sociological Review, 25*(1), 9–21.

Livingstone, S. (2002). *Young people and new media.* London: Sage.

Livingstone, S. (2009). *Children and the internet. Great expectations, challenging realities.* Cambridge: Polity.

Luijkx, K., Peek, S., & Wouters, E. (2015). "Grandma, you should do it—it's cool": Older adults and the role of family members in their acceptance of technology. *International Journal of Environmental Research and Public Health, 12*(12), 15470–15485.

Mandić, S. (2001). Residential mobility versus 'in-place' adjustments in Slovenia: Viewpoint from a society 'in transition'. *Housing Studies, 16*(1), 53–73.

Moffatt, K., David, J., & Baecker, R. M. (2013). Connecting grandparents and grandchildren. In C. Neustaedter, T. Harrison, & A. Sellen (Eds.), *Connecting families: The impact of new communication technologies on domestic life* (pp. 173–193). London: Springer.

Morgan, D. (2011). *Rethinking family practices.* Dordrecht: Springer.

Mori, K., & Harada, E. T. (2010). Is learning a family matter?: Experimental study of the influence of social environment on learning by older adults in the use of mobile phones. *Japanese Psychological Research, 52*(3), 244–255.

Neustaedter, C., Harrison. T., & Sellen, A. (Eds.) (2013). *Connecting families: The impact of new communication technologies on domestic life.* Dordrecht: Springer.

Oksman, V., & Turtiainen, J. (2004). Mobile communication as a social stage: Meanings of mobile communication in everyday life among teenagers in Finland. *New Media & Society, 6*(3), 319–339.

Oláh, L. S. (2015, May). Changing families in the European Union: Trends and policy implications. In *United Nations Expert Group Meeting "Family policy development: Achievements and challenges" in New York, May* (pp. 14–15).

Olivier, P., & Wallace, J. (2009). Digital technologies and the emotional family. *International Journal of Human-Computer Studies, 67*(2), 204–214.

Piumatti, G., Garro, M., Pipitone, L., Di Vita, A. M., & Rabaglietti, E. (2016). north/south differences among italian emerging adults regarding criteria deemed important for adulthood and life satisfaction. *Europe's Journal of Psychology, 12*(2), 271–287.

Rainie, L., & Wellman, B. (2012). *Networked: The new social operating system.* Cambridge, MA: MIT Press.

Shove, E., Pantzar, M., & Watson, M. (2012). *The dynamics of social practice: Everyday life and how it changes.* London: Sage.

Siibak, A., & Tamme, V. (2013). 'Who introduced granny to Facebook?': An exploration of everyday family interactions in web-based communication environments. *Northern Lights: Film & media studies Yearbook, 11*(1), 71–89.

Singer, D. G., & Singer, J. L. (2012). *Handbook of children and the media.* London: Sage.

Taipale, S., Petrovčič, A., & Dolničar, V. (2018). Intergenerational solidarity and ICT usage: Empirical insights from Finnish and Slovenian families. In S. Taipale, T.-A. Wilska, & C. Gilleard (Eds.), *Digital technologies and generational identity: ICT usage across the life course* (pp. 68–86). London & New York, NY: Routledge.

Taipale, S., Wilska, T.-A., & Gilleard, C. (Eds.). (2018b). *Digital technologies and generational identity: ICT usage across the life course.* London & New York, NY: Routledge.

Tsai, T. H., Ho, Y. L., & Tseng, K. (2011). An investigation into the social network between three generations in a household: bridging the interrogational gaps between the senior and the youth. In

Online Communities and Social Computing (pp. 277–286). 4th International Conference, OCSC 2011, Dordrecht: Springer.

Venkatesh, A., Dunkle, D., & Wortman, A. (2011). Family life, children and the feminization of computing. In R. Harper (Ed.), *The connected home: The future of domestic life* (pp. 59–76). Dordrecht: Springer.

Venkatesh, A., Kruse, E., & Shih, E. (2003). The networked home: An analysis of current development and future trends. *Cognition, Technology & Work, 5*(1), 23–32.

Vincent, J. (2006). Emotional attachment and mobile phones. *Knowledge, Technology & Policy, 19*(1), 39–44.

Vincent, J., & Fortunati, L. (Eds.). (2009). *Electronic emotion, the mediation of emotion via information and communication technologies*. Bern: Peter Lang.

Chapter 3
Technological Drivers and Sociocultural Change

Abstract In this chapter, the suggestion is made that the relevance and timeliness of the concept of digital family is owing to certain advances in personal communication technologies that, together with wider social changes in the developed world, facilitate people's participation in a digital society. First, we take a look at recent technological advancements in personal communication technology and social media applications, after which we examine more in-depth some of the major sociocultural transformations to have paved way for the emergence and rise of the digital family.

Keywords Connected home · Digital skills · Family roles · Infrastructure · Internet history · Mobile communication · Personal communication

The rise of the digital family has been propelled by a number of parallel forces impacting different social groups differently and moving at a different pace from one country to the next. Among these forces are processes of technological advancement and sociocultural development, most usefully looked at as expressions of the kind of global trends studied in-depth by Castells (2010). According to the latter, new technologies such as microprocessors, Internet technology, telecommunications networks and genetic engineering have, from the 1960s onwards, been adding up to a new social morphology of networks around which communication and power relationships organize themselves across countries. It is, claims Castells (2010), the forms and compositions of families that, under the joint impact of the new technologies of reproduction and the crisis of patriarchalism, have been reshaped in the course of the past half-century, at the same time as networking technologies have only more and more begun to influence the way family members relate to one another and experience family connection. Since these large-scale technological and sociocultural changes have been well documented by not only Castells but also others (e.g. van Dijk, 2012), this chapter provides only a brief overview of the wide and multifaceted developments in question, concentrating on recent advancements in mobile communications and one country case, Finland. In what follows, I will first take a look at technological advances in the field of personal communication technologies and social media, before examining a little more in-depth some of the major sociocultural transformations that paved way for the emergence and rise of the digital family.

© Springer Nature Switzerland AG 2019

S. Taipale, *Intergenerational Connections in Digital Families*,

https://doi.org/10.1007/978-3-030-11947-8_3

Tools, Connections, Content

For many people, the smartphone is the most concrete reflection of the fast pace of technological advancement. Due to its limited internal memory or processor capacity, or steadily weakening battery, it grows old in just a couple years, reminding us about how rapidly everything in mobile communication technology keeps changing around us. In retrospect, it may seem like the shift from the basic feature phones to Internet-enabled smartphones was swift and easy. While, in fact, it took several years to complete, the process was carried through incrementally, through a series of small steps and involving no major compatibility problems between the old and the new versions of the technology. Some may, however, still recall the period as that of competing network standards (CDMA/GSM/LTE) and locked SIM cards that prevented the use of certain devices and subscription plans outside of their home domain, in other countries and continents.

For mobile device users who regularly switched to newer models as these became available, technological advances both in mobile hardware and software were, indeed, relatively easy and effortless to adopt into use. The previously small screen become gradually bigger, the dialling keyboard was replaced first by a full-sized QWERTY keyboard, later a virtual keypad. Simultaneously, the mobile phones became more and more closely linked with the Internet, beginning with the Wireless Application Protocol (WAP) and General Packet Radio Service (GPRS), all the way to the recent 3G and today's 4G technology. The mobile phone turned into the smartphone, or, indeed, just 'the phone', as, parallel with this process, landline telephones became increasingly rare.

From the perspective of digital families, it was of particular significance that the improvements in mobile communications *tools* and *infrastructure* enabled a potentially wider reach of people. Soon enough, in Europe and in many other parts of the developed world, family members of all ages had gradually become users of mobile phones, at least of basic feature phones. In families, the technological advances have meant a possibility to transmit larger amounts of data and to extend the modes of communication from voice-only and text-only to also include images and teleconferencing. In extended families, however, the asynchronous uptake of newest technologies and applications has also translated into new mechanisms of social inclusion and exclusion. Depending on their motivation, needs and desire, family members have voluntarily opted either in or out of using certain mobile communication applications or social media tools, but sometimes they can also be purposely kept outside the online family circle, such as when children want to avoid their parents' watchful eyes on social media platforms.

Connections through faster mobile broadband technology have also meant new technological affordances[1] compared to the feature phone, fixed broadband con-

---

[1]The concept of affordance, originally coined by the perceptual phycologist Gibson (1979) and later developed by Norman (1988) in the field of human–computer interaction studies, refers to the intended and unintended uses of a technology that unfold to people as they interact with technologies in a certain environment.

nections and wireless Internet hotspots. What the smartphone offered for its users was both new asynchronous (email, voice messaging services, etc.) and synchronous (chats, video calls, etc.) modes of communication. These made it possible to select the most appropriate communication tools for each family member, taking into account the individually differing skills levels and preferences (cf. Taipale & Farinosi, 2018). Thanks to high-quality built-in cameras, smartphones also took the aspect of visuality in family communication to a completely new level. In many families, the exchange of images and videos quickly became part of people's daily routine of keeping in touch with one another, complementing the text- and voice-based communication. In extended families, regular exchange of small messages has subsequently become essential for social connectivity and the family members' continual sense of togetherness (Chap. 7).

As the history of technology has repeatedly demonstrated, a new technological innovation only seldom replaces its predecessors completely. This was the case also with mobile broadband technology, which has not fully substituted fixed broadband connections, laptop computing and desktop computers at homes and workplaces. Fixed Internet connections have persisted as a fast and often more reliable way to connect to the Internet than mobile net subscriptions. In consequence, people often draw upon a large selection of personal technologies instead of employing just one type. At times, a communicative act may even call for its successful accomplishment the use of many communication technologies providing different affordances in a consecutive manner. For example, an email may first be read on a mobile screen, then the file attachment is printed out using a separate printer, while, finally, a reply is sent from one's desktop computer (Fortunati & Taipale, 2017).

To better grasp this ever-expanding catalogue of personal media technologies, Madianou and Miller (2011, 2012) have coined the term polymedia. In a situation of wide accessibility and low-cost barriers of contemporary communication technologies, the notion is intended to help clarify the social, moral and political consequences of our choices when selecting from among the available technologies in different social contexts. High adoption rates and wide accessibility do not mean that geographical inequalities in network coverage or the costs of communication would have become inconsequential for existing and would-be technology users. There are still remote and sparsely populated areas in all developed countries where both fixed broadband and high-speed mobile networks remain unavailable, due to the low number of (potential) paying customers. Because of this, to be sure, many practical measures have been introduced, for instance, within the European Union, to improve accessibility and increase competition all across the continent, with political decisions taken to lower the prices and make the networks more uniform and predictable to their users (e.g. the European 'Roam Like at Home' regulation of 15 June 2017).

The switch from feature phones to smartphones and mobile Internet applications has also had consequences for the actual and potential reach of the communication (see Baym, 2015; Gurak, 2003; Taipale & Farinosi, 2018). Early mobile phone communication was mostly about voice calls and text messaging, with especially young people embracing the two. As concerns their reach, however, the new means were still largely restricted to one-to-one communication. It was only with Internet-enabled smartphones that the possibility to reach several people or large groups at once became real: instant messengers and social networking applications made one-to-many and many-to-many communication finally possible. For distributed extended families, this opened up an entirely new world of possibilities, enabling one to reach all family members at once with the help of the smartphone.

In addition to the new tools and infrastructures, there was also a third major technological driver behind the developments, one that had to do with the *contents* of communication. Compared to feature phones, smartphones incorporate into themselves a far greater number of old-media features, enabling good quality access to television, radio, camera and newspaper functions (Fortunati & Taipale, 2017). In that sense, the smartphone is an example of technological convergence. Moreover, while feature phones were decorated with covers, stickers and mobile jewellery on the outside, smartphones are personalizable and customizable also inside. Apart from enabling the personalization of the interface to the core, however, also the contents availed by the smartphone are personalized, based on a multiplicity of algorithms that track users' prior online behaviour to decide what information the applications will display and what they will not.

Within families, this internal personalization of smartphones has at least two main implications. First of all, family communication may become compartmentalized according to the modes of interaction and applications, such as when children decide to use a certain application with parents and another one with siblings and friends (Hänninen, Taipale, & Korhonen, 2018). Second, families can also actively seek common platforms for their family communication, ones that meet the different communicative preferences and styles of their members. Instant messengers like WhatsApp have shown themselves to be quite adaptable in this regard (Rosales & Fernández-Ardèvol, 2016; Siibak & Tamme, 2013; Taipale & Farinosi, 2018).

Overall, advances in mobile communication and broadband technology have resulted in a new technical basis on which extended families can manage multiple family ties, both on the move and when stationary. While smartphones as a multipurpose communication tool are emblematic of this development, they have not, however, fully replaced all other forms of communication and communication tools, such as laptops and desktop computers, which continue to play an important role in family communication. Moreover, with the ever-widening spectrum of personal and household technologies that require constant maintenance, updating and reconfiguration, families have also been forced to take on new household duties and practices related to the maintenance of the digital home (for more on this, see Chap. 6).

The Changing Home, Skills and Family Roles

While technological advances have provided the new tools, infrastructures and contents necessary for families to be able to become 'digital ready', much has also needed to happen to the family members themselves, their respective roles and their home equipment to enable the transition. In this section, I take a brief look at some of the major developments impacting the domestic sphere, developments that have effected a reconfiguration of family relationships and roles and improved families' abilities to stay connected via digital media and digital communication tools. First, evidence of the swift digitalization of homes is presented, concentrating on the case of Finland. After that, the focus is turned to certain changes in people's digital skills, in particular those that contribute to individuals' ability to use new technologies for the benefit of family communication. The section then ends with a concise overview of previous research on the redistribution of family roles, especially in connection to the management of digital technologies at home.

The Equipped and Connected Home

Like individuals, also the homes as places have quickly become more connected and equipped with digital ICTs. For example, in just 2 years, from 2014 to 2016, the share of Finnish households owning smartphones went up from 69 to 82%, while those with a wide-screen television in them went up from 81 to 86%, a smart television from 19 to 30%, at least one tablet computer from 39 to 56%, and a wireless LAN connection from 54 to 66%. Some older technologies, like desktop computers, digital cameras and printers, had already reached a saturation point by 2014, with no significant increases in ownership levels after that (Statistics Finland, 2017)

Despite their quick diffusion, however, the distribution of even the most widely adopted digital technologies has not been uniform within families and across age groups. The mobile phone provides a case in point. Today, it an almost universal communication tool owned by virtually everyone in families. In Finland, all persons under 75 years used a mobile phone in 2017, and among those in the oldest age category, or 75–89 years, the share of mobile phone users was as high as 91% that same year (Statistics Finland, 2017). All the same, the age gap was still very prominent when it came to smartphone ownership. Among all those aged 65 or under, the smartphones were already very common in 2017, with 77–99% of all those surveyed owning one, but in the older categories ownership was less common: among those aged 65–74 and 75–89 years, the figures were 49 and 15%, respectively (Statistics Finland, 2017).

The data from Finland appears to confirm that the size of household matters when it comes to digital media and communication technology ownership. Compared to single-person households, Finnish households with three or more members have clearly more often in them at least some kind of computer (99% of them did in 2017, compared to 73% of single-person households), at least one laptop computer (89% vs. 55%), a desktop computer (48% vs. 22%), a tablet computer (82% vs. 27%), an Internet connection (99% vs. 77%), and an in-house WLAN network (82% vs. 33%) (Statistics Finland, 2017). Also, studies from other countries have shown the size of household to be linked to the degree of accumulation of digital technologies. In

Table 3.1 Share of Finnish households with computer (any), by net household income and location, 2013–2017 (in percentage points)

	2013	2014	2015	2016	2017
Net income, €/month					
2100 or less	61	65	66	71	75
2099–3099	86	88	91	94	99
3100–5099	97	98	98	98	99
5100 or more	99	99	99	99	100
Capital region	87	87	88	89	93
Major cities	84	82	82	87	89
Other cities and towns	80	80	83	85	87
Small and rural municipalities	75	76	77	79	82
All households	81	81	82	85	87

Source Statistics Finland (2017)

general, households with children appear to outperform all other types of households in this regard (Venkatesh, Dunkle, & Wortman, 2011).

As Table 3.1 shows, the share of Finnish households with at least one computer of any kind continued to rise slowly between 2013 and 2017. During this period, moreover, low-income households caught up wealthier household. At the same time, the speed of household computerization remained almost the same in urban and rural areas, being only slightly lower in small and rural municipalities. In this particular connection, it is worth noting, however, that households in urban and rural areas, just as low and high-income households, differed from each other in terms of their composition. Higher income households typically had more (two) income earners, and in rural areas the shares of those under 15 and those aged 65 or older were larger than in the capital region and in major cities (Statistics Finland, 2017).

Table 3.2, for its part, shows the share of Finnish households connected to the Internet through a mobile or fixed-line broadband subscription. Between 2013 and 2017, this share went up slowly, from 81 to 88%. During the period, the gap between the high and the low-income households narrowed considerably. With the exception of households with a net income of 2100 euros or less a month, many of which can be presumed to have included low-income pensioners, all households in the country had access to the Internet by the end of 2017. Households in rural areas remained somewhat less connected than those in the cities and towns, presumably, again, at least in part due to the relatively higher presence of older people in the countryside.

Apart from acquiring a greater variety of equipment and appliances and becoming better connected to the Internet, the digitalization of homes also comes about through

Table 3.2 Share of Finnish households with Internet connection (any), by net household income and location, 2013–2017 (in percentage points)

	2013	2014	2015	2016	2017
Net income, €/month					
2100 or less	61	65	76	72	77
2099–3099	87	89	91	93	94
3100–5099	97	98	98	99	100
5100 or more	99	99	99	99	100
Capital region	88	87	89	91	93
Major cities	84	84	82	86	90
Other cities and towns	80	80	84	86	88
Small and rural municipalities	77	76	77	78	83
All households	81	81	82	85	88

Source Statistics Finland (2017)

the sheer numerical accumulation of same equipment/appliances and services. New digital devices are increasingly intended for personal usage, which means that having just one unit of each kind in the household is typically not enough (except in one-person households, of course). One indication of such accumulation is the increase in the number of Internet subscriptions per household. Table 3.3 shows the development in Finland in this regard, in the time period 2012 through 2015. Assuming that the overall relative distribution of households of different sizes did not radically change during this period, one can observe that more and more households acquired multiple Internet connections in the course of it. In just 4 years, the share of households with only one Internet subscription halved, from 41 to 20%. At the same time, the share of households with three or more subscriptions doubled, from 21 to 45%. The overall trend was the same in both rural and urban areas, although households in small and rural municipalities made considerable progress in catching up with others.

Parallel to the increase in the number of digital communication tools and the expansion of the Internet connectivity, more and more people have begun to consider new media and ICTs as necessities in life. As Venkatesh et al. (2011, p. 61) found, the share of US Americans claiming themselves unable to imagine life without a home computer increased from 44 to 61% between 1999 and 2010. Also the share of those considering computers to have made it easier for them to organize their family and social life also increased, from 34 to 43%, between 2003 and 2010. Along the same lines, Wilska and Kuoppamäki (2018) found the number of Finns viewing personal computers, access to the Internet and mobile phones to constitute a necessity in life to have increased steadily between 1999 and 2014. During the same period, the differences between age cohorts in the perceived necessity of ICTs became smaller, apart from the pre-existing gap between pre and post-World War II generations that

Table 3.3 Number of Internet subscriptions per household in Finland, by location, 2012, 2014 and 2015 (in percentage points)

	2012			2014			2015		
	Only one	Two or more	Three or more	Only one	Two or more	Three or more	Only one	Two or more	Three or more
Capital region	34	33	26	17	26	44	17	22	49
Major cities	40	28	19	19	29	35	18	28	35
Other cities and towns	39	23	21	21	20	38	20	21	42
Small and rural municipalities	48	18	18	23	19	33	21	16	39
All households (persons aged 16–74)	41	25	21	21	28	40	20	25	45

Source Statistics Finland (2017)

kept widening. All in all, the homes, at least in Finland, have thus become well equipped and ready for the digital families to begin inhabiting them. Yet, without elaborating on the question further, we may also assume country differences to likely be notable in this regard, too.

Improvements in Digital Skills

Digital skills are required if one is to make good use of new media and communication technologies in everyday life (Martínez-Cantos, 2017). The literature in the field is replete with concepts geared to identifying and describing the phenomenon at stake and what is novel in it (e.g. 'new media literacy', 'digital literacy', 'ICT skills'). At the most general level, digital skills can be divided into operational skills needed to use a range of technologies for social and creative purposes, and strategic skills, required to understand the social and commercial risks and opportunities involved (see Helsper & Eynon, 2013). Following van Deursen and van Dijk (2015), one can also make a further distinction between medium-related skills and content-related skills. In extended families, a cross-generational increase in family members' digital skills can be considered as one prerequisite for better intergenerational communication via new media and communication technologies.

While much is written about digital skills on a general plane, longitudinal studies on the development of skills over time and across age groups are still lacking. Yet, the perspective that kind of research could offer would be particularly useful to understanding the rise of digital families. As van Deursen and van Dijk (2015), for example, have shown, operational and formal Internet skills of those aged 18–65 in the Netherlands increased evenly across the entire age spectrum over the time period 2010–2013. The reason why those skills matter is that they are prerequisites for the acquisition of higher level skills related to informational and strategic Internet skills. In terms of their information seeking skills, as van Deursen and van Dijk (2015) also found, Dutch respondents aged 65 or older had by 2013 caught up with their younger counterparts. In another study, van Deursen, van Dijk, and Peters (2012) could, furthermore, note that, when older people attained a certain level of operational and formal Internet proficiency, they also managed to translate these into higher level informational and strategic Internet skills.

Fortunati, Taipale, and de Luca (2017), for their part, investigated possible changes in self-reported ICT skills in five European countries (France, Germany, Italy, Spain and UK) in the time period 1996–2009. Given the study period involved (going back in time even more than two decades), their findings concerned rather basic ICT skills in using significantly more primitive technology than what people generally use today (relatively early-generation personal computers). The study found those claiming themselves not to know how to use a personal computer to have slightly decreased over the time period, while in the youngest age groups, those aged 14–17 and 18–24, the share of respondents considering themselves as 'Expert at using computers' slightly increased. However, even among all those aged 25 or over, the total share of respondents stating that 'I can get by' increased. This was, moreover, especially prominent in the age group 65 years or older.

Research has also sought to investigate any changes in the digital skills gap between European men and women. In one study, Martínez-Cantos (2017), scrutinizing Eurostat data on digital economy and society, found that, while both men's and women's digital skills improved between 2017 and 2013, gender differences in this regard did not vanish completely. Among those with less advanced digital skills, the gender gap remained pronounced in the oldest and less educated groups. Interestingly, when it came to those with higher level digital skills, gender differences were, instead, most prominent among middle-aged and younger respondents. On the other hand, earlier studies have found that, even though women's self-perception of their digital skills is comparatively poorer than that of men, there are no significant differences in the two genders' actual capabilities of using the Internet (van Deursen & van Dijk, 2015).

In addition to gender and age, education appears to be another major social-demographic factor connected to individuals' digital skills levels. It has even been suggested that, globally speaking, the level of education is perhaps the most consistent determinant of Internet skills levels: the higher one's level of education, the more capable one is in using the Internet for various purposes (van Deursen & van Dijk, 2011, 2015). The findings form a large-scale international adult-skills survey by the OECD point in the same direction, showing higher levels of education to be connected to, among other things, greater proficiency in problem-solving in technology-rich environments more in general (OECD, 2013; see also Hämäläinen, De Wever, Malin, & Cincinnato, 2015). What is, however, alarming in this regard is that while Internet skills have increased overall, at the same time also the gap between those with more education and those with less education has tended to increase. Indeed, that trend has been so solid that the inequality it expresses has been predicted to remain a long-lasting feature of the ongoing developments (van Deursen & van Dijk, 2015).

What we may then conclude is that one notable result from the generally positive development of digital skills is that the so-called *age divide* has grown old. The largest age gap in ICT adoption, use and skills are now between 'old' and 'oldest old' groups (Friemel, 2016; Hargittai & Dobransky, 2017; Petrovčič, Slavec, & Dolničar, 2018). Many studies confirm the primary reason for digital disengagement among older users to no longer be lack of access or equipment, but, rather, lack of skills and, even more often, lack of interest and self-confidence in ICT matters (van Deursen & Helsper, 2015; Siren & Knudsen, 2017; van Deursen & van Dijk, 2015).

In short, research shows digital skills to have improved across all age groups, almost regardless of the measure used. As for the developments affecting families more specifically, of paramount importance has been that older people have gained more skills and self-confidence, leading many of them to begin using ICTs as an integral part of their daily life (e.g. Khvorostianov, 2016). In practice, the improved digital skills have enabled parents and grandparents to better gear up for, and put to use in their everyday life, the same communication and media technologies that their children and grandchildren use (e.g. Siren & Knudsen, 2017). Moreover, for family connectivity, it has been crucial that the share of those able to 'get by' with new technologies has grown, even if at the same time certain skills gaps between younger and older users and between more and less educated are likely to persist. A positive attitude shown in the family towards technology ensures a fertile ground for further digital skills learning. As some studies have already been able to observe, the family seems to be a more important place for adolescents to learn ICT use than the school, as at home they are freer to explore devices and programmes, and parents can provide more direct and personal support than a teacher when difficulties emerge (Zhong, 2011). Similarly, it can be expected that a supportive, pro-technology atmosphere in the home encourages also older and less-skilled family members' uptake of, and experimentation with, new technologies.

Changes in Family Roles

Classical sociological theories stress parents' central role as agents of socialization in families (e.g. Parsons, 1951). Parents' influence on the norms and values that children adopt and on the way they act is at its strongest during the first few years of the young person's life. The influence of peers, siblings and the media begins to gradually grow and become more pronounced as children grow older. More recently, family studies have begun to also pay more attention to two-way influences between older and younger family members (e.g. Kuczynski & De Mol, 2015). Although parents and children are differently positioned within the line of family generations, both are involved in the shaping of their own and others family members' everyday relationships (e.g. Alanen, 2009, p. 161). This continuous shaping of family roles forms part of the wider process of 'doing family' (Chap. 2).

Already from early on, studies on changing family roles in digital technology adoption and usage pointed to a reversal of traditional roles in these areas: children, it was found, had become teachers for their parents in technical matters. In one study of computer help-seeking among 93 US families, for instance, it was teenagers rather than they parents who most often provided technical help and know-how to others in the family (Kiesler, Zdaniuk, Lundmark, & Kraut, 2000). Also, research in the United Kingdom has found traditional adult–child relationships in many households to have been reversed in this regard, with the children possessing more technological competence than their parents (Holloway & Valentine, 2003). More in general, children have also been found to be far more able than older family members to define the meaning and uses of the computer in the home (Facer, Furlong, Furlong, & Sutherland, 2003).

In distributed and extended digital families, family relationships and roles are constantly reproduced in and through intergenerational practices that increasingly more are related to, or mediated by, digital technologies. Nevertheless, many researchers still tend to choose just either parents or children as their point of departure in their research. There is, for instance, a large body of research on the ways in which parents guide and mediate their children's use of digital communication technologies and media. Focusing mainly on parents' role as educators and caretakers, such studies have concentrated on the kind of parenting styles that facilitate children's healthy and risk-free use of new media (e.g. Clark, 2013; Livingstone & Haddon, 2009; Livingstone et al., 2015; Wartella et al., 2013). Another branch of research has proceeded from an opposite standpoint, looking at children's impact on their parents' media consumption and media usage behaviour (e.g. Correa, 2014; Correa, Straubhaar, Spence, & Chen, 2015; Eynon & Helsper, 2015). While the two approaches, separately and jointly, have brought to light many interesting aspects of children's comparatively high level of agency vis-à-vis their parents, and highlighted parents' strategies of using new technology to monitor and support their children, they have nevertheless largely failed to adequately address the two-way child–parent influences.

Indeed, studies combining the two viewpoints are not many. One of few doing so is a recent study by Nelissen and Van den Bulk (2018) that investigated the nature of child–parent interactions around new technology use in 187 cases from Belgium. In the study, also the children (and not just the parents) saw themselves (the children) as active agents teaching digital media use to their parents. Furthermore, the parents and the children reported also the extent of the technological guidance provided by the latter identically. In this study, the child–parent digital media guidance mainly involved the use of personal mobile media tools, such as smartphones, tablets and their applications, but not computers, computer programmes, email and online purchases. What this may imply is that children's ability to guide and assist their parents might, to a larger extent than researchers have thus far understood, be limited to those devices and applications that children themselves use actively. Adulthood entails tasks involving the use of devices, programmes and solutions that children either use not at all or are not familiar with (e.g. professional computer programmes, online banking or e-government services), and thus, from the parents' perspective, this circumstance might then not always allow them to receive the amount of assistance they need.

The gender issue discussed above also applies to ICT guidance and help provision in families, although, over time in many developed countries, the gendered patterns in ICT adoption and use in families have become considerably less distinct, sometimes even completely disappearing (see, e.g. Plowman, McPake, & Stephen, 2010; Rideout & Hamel, 2006; Venkatesh, Dunkle, & Wortman, 2011). Nevertheless, there still seem to be some specific gender differences left, also pertaining to the reversed family roles. Correa's (2014) study, for instance, suggests that women's media and technology use might be more influenced than men's by their children's guidance. Given this finding, it is then interesting that, for instance, in Nelissen and Van den Bulck's (2018) study media-related family conflicts were reported similarly by both men and women, by both parents and their children.

In this connection, it is worth stressing that any possible or real changes in family roles, whether related to new technology or other things in family life, do not, however, mean the end of child–parent conflicts. Quite the contrary: already more than a decade ago, Mesch (2006) noted how in fact more media-related family conflicts were reported in families where teenagers' digital expertise was perceived as greater than that of their parents. More recently Nelissen and Van den Bulck (2018) have confirmed this finding, discovering that, in families where there was much child–parent guidance around digital media use, more family conflicts were reported. Indeed, that children's more advanced skills in ICT use often cause discomfort in parents and can hence foster family conflicts was already noticed some twenty years ago (e.g. Kiesler et al., 2000; Watt & White, 1999). However, what makes technology-related conflicts different from other family conflicts is the fact that especially older parents lack a reference model for how to deal with them (see Plowman, McPake, & Stephen, 2010). Since digital technologies arrived rather late in their lives, the parents of today's children, when they were young, never themselves experienced comparable conflicts with their own parents about, say, excessive screen time or playing games not suitable for children.

Moreover, the extent to which technology and media-related conflicts occur in families is likely to vary across countries, depending on the prevailing family culture and values. Nevertheless, previous research suggests that the spread of digital technologies may have led to more democratic, and perhaps more intimate, child–parent relationships, at least in Europe (e.g. Livingstone, 1998). In societies with a more conservative family culture (as, for instance, in Asia), the use of new communication technologies may have, in addition, led to gendered practices being challenged and family hierarchies becoming less rigid (Lim & Soon, 2010). On the other hand, however, research has also indicated that, in the presence of an authoritarian family culture, expressions of sentiments in technology-related interactions are less likely (Cardoso, Espanha, & Lapa, 2012; Haddon, 2009).

In general, to be sure, the above-presented reconfigurations of family roles ascribable to ICT usage seem to co-occur with wider and more comprehensive changes in family values. In Europe, traditional child–parent power hierarchies appear on the whole to be slowly eroding. As the 2018 European Values Study reveals, Europeans

today in general consider family as something either very important or quite important in their lives. Yet, the differences from country to country in this regard remain
notable, as observed earlier. While in Finland 73% of those surveyed agreed with the
statement 'adult children have a life of their own and should not be asked to sacrifice
their own well-being for the sake of their parents', only 39% of respondents in Slovenia and 21% in Italy gave the same answer (European Values Study, 2018). At the
same time, however, other studies have shown Europeans to increasingly value children's well-being over that of their parents. The change, to be sure, has been slower
in Germany than in, say, the Netherlands, Sweden, France or the United Kingdom, as
measured for the time period 1990–2008 (Ivan, Da Roit, & Knijn, 2014). Overall, as
Park and Lau (2016) have showed, child independence tends to be appreciated more
in wealthy nations and among highly educated people, with child obedience being
the stronger value in poorer societies with characteristically lower levels of education. Having said that, it nonetheless appears likely that children on the whole have
gained more power to influence decision-making in their families when it comes to
technology purchases and use. Yet, given the existing differences in family structure
and perceptions about who belongs to the family, it seems plausible that this new
power and the new family roles associated with it are differently manifested in, not
least, the three countries involved in this particular study.

References

Alanen, L. (2009). Generational order. In J. Qvortrup, W. A. Corsaro, & M.-S. Honig (Eds.), *The
palgrave handbook of childhood studies* (pp. 159–174). Basingstoke, UK: Palgrave Macmillan.
Baym, N. K. (2015). *Personal connections in the digital age* (2nd ed.). Cambridge: Polity.
Cardoso, G., Espanha, R., & Lapa, T. (2012). Family dynamics and mediation: Children, autonomy
and control. In E. Loos, L. Haddon, & E. Mante-Meijer (Eds.), *Generational use of new media*
(pp. 49–70). London & New York, NY: Routledge.
Castells, M. (2010). Rise of the network society. In *The information age: Economy, society, and
culture* (2nd ed., Vol. 1). Malden, MA: Wiley.
Clark, L. S. (2013). *The parent app: Understanding families in the digital age*. Oxford: Oxford
University Press.
Correa, T. (2014). Bottom-up technology transmission within families: Exploring how youths influ-
ence their parents' digital media use with dyadic data. *Journal of Communication, 64*(1), 103–124.
Correa, T., Straubhaar, J. D., Spence, J., & Chen, W. (2015). Brokering new technologies: The role
of children in their parents' usage of the internet. *New Media & Society, 17*(4), 483–500.
European Values Study. (2018). *Atlas of European values*. Retrieved from http://www.
atlasofeuropeanvalues.eu/new/home.php?lang=en.
Eynon, R., & Helsper, E. (2015). Family dynamics and internet use in Britain: What role do children
play in adults' engagement with the Internet? *Information, Communication & Society, 18*(2),
156–171.
Facer, K., Furlong, J., Furlong, R., & Sutherland, R. (2003). *Screenplay: Children and computing
in the home*. New York, NY: Routledge.
Fortunati, L., & Taipale, S. (2017). Mobilities and the network of personal technologies: Refining
the understanding of mobility structure. *Telematics and Informatics, 34*(2), 560–568.
Fortunati, L., Taipale, S., & de Luca, F. (2017). Digital generations, but not as we know them.
Convergence. Advance online publication. https://doi.org/10.1177/1354856517692309.

Friemel, T. N. (2016). The digital divide has grown old: Determinants of a digital divide among seniors. *New Media & Society, 18*(2), 313–331.

Gurak, L. J. (2003). *Cyberliteracy: Navigating the internet with awareness*. New Haven, CT: Yale University Press.

Hämäläinen, R., De Wever, B., Malin, A., & Cincinnato, S. (2015). Education and working life: VET adults' problem-solving skills in technology-rich environments. *Computers & Education, 88*, 38–47.

Hänninen, R., Taipale, S., &Korhonen, A. (2018). Refamilisation in the broadband society. The effects of ICTs on family solidarity in Finland. *Journal of Family Studies.* Advance online publication. doi.org/10.1080/13229400.2018.1515101.

Hargittai, E., & Dobransky, K. (2017). Old dogs, new clicks: Digital inequality in skills and uses among older adults. *Canadian Journal of Communication, 42*(2), 195–212.

Helsper, E. J., & Eynon, R. (2013). Distinct skill pathways to digital engagement. *European Journal of Communication, 28*(6), 696–713.

Holloway, S. L., & Valentine, G. (2003). *Cyberkids: Children in the information age*. Psychology Press.

Ivan, G., Da Roit, B., & Knijn, T. (2014). Children first? Changing attitudes toward the primacy of children in five European countries. *Journal of Family Issues, 36*(4), 1982–2001.

Khvorostianov, N. (2016). "Thanks to the internet, we remain a family": ICT domestication by elderly immigrants and their families in Israel. *Journal of Family Communication, 16*(4), 355–368.

Kiesler, S., Zdaniuk, B., Lundmark, V., & Kraut, R. (2000). Troubles with the internet: The dynamics of help at home. *Human-Computer Interaction, 15*(4), 323–351.

Kuczynski, L., & De Mol, J. (2015). Dialectical models of socialization. In W. F. Overton, & P. C. M. Molenaar (Eds.), *Theory and method. Volume 1 of the handbook of child psychology and developmental science* (pp. 326–368). Hoboken, NJ: Wiley.

Lim, S. S., & Soon, C. (2010). The influence of social and cultural factors on mothers' domestication of household ICTs—Experiences of Chinese and Korean women. *Telematics and Informatics, 27*(3), 205–216.

Livingstone, S., & Haddon, L. (2009). *EU Kids Online: Final report*. London: LSE. Retrieved from http://www.lse.ac.uk/media@lse/research/EUKidsOnline/EU%20Kids%20I%20(2006-9)/EU%20Kids%20Online%20I%20Reports/EUKidsOnlineFinalReport.pdf.

Livingstone, S. (1998). Mediated childhoods: A comparative approach to young people's changing media environment in Europe. *European Journal of Communication, 13*(4), 435–456.

Livingstone, S., Mascheroni, G., Dreier, M., Chaudron, S., & Lagae, K. (2015). *How parents of young children manage digital devices at home: The role of income, education and parental style*. LSE: London. Retrieved from http://www.lse.ac.uk/media@lse/research/EUKidsOnline/Home.aspx.

Madianou, M., & Miller, D. (2011). Mobile phone parenting: Reconfiguring relationships between Filipina migrant mothers and their left-behind children. *New Media & Society, 13*(3), 457–470.

Madianou, M., & Miller, D. (2012). *Migration and new media: Transnational families and polymedia*. London & New York, NY: Routledge.

Martínez-Cantos, J. L. (2017). Digital skills gaps: A pending subject for gender digital inclusion in the European Union. *European Journal of Communication, 32*(5), 419–438.

Mesch, G. S. (2006). Family relations and the Internet: Exploring a family boundaries approach. *The Journal of Family Communication, 6*(2), 119–138.

Nelissen, S., & Van den Bulck, J. (2018). When digital natives instruct digital immigrants: Active guidance of parental media use by children and conflict in the family. *Information, Communication & Society, 21*(3), 375–387.

OECD (2013) *Country note survey of adults skills. First results. Finland*. Retrieved from http://www.oecd.org/skills/piaac/Country%20note%20-%20Finland.pdf.

Park, H., & Lau, A. S. (2016). Socioeconomic status and parenting priorities: Child independence and obedience around the world. *Journal of Marriage and Family, 78*(1), 43–59.

Parsons, T. (1951). *The social system*. New York, NY: Free Press.

Petrovčič, A., Slavec, A., & Dolničar, V. (2018). The ten shades of silver: Segmentation of older adults in the mobile phone market. *International Journal of Human-Computer Interaction, 34*(9), 845–860.

Plowman, L., McPake, J., & Stephen, C. (2010). The technologisation of childhood? Young children and technology in the home. *Children and Society, 24*(1), 63–74.

Rideout, V., & Hamel, E. (2006). *The media family: Electronic media in the lives of infants, toddlers, preschoolers and their parents.* Henry J. Kaiser Family Foundation. Retrieved from https://kaiserfamilyfoundation.files.wordpress.com/2013/01/7500.pdf.

Rosales, A., & Fernández-Ardèvol, M. (2016). Beyond WhatsApp: older people and smartphones. *Revista Română de Comunicare şi Relaţii Publice, 18*(1), 27–47.

Siren, A., & Knudsen, S. G. (2017). Older adults and emerging digital service delivery: A mixed methods study on information and communications technology use, skills, and attitudes. *Journal of Aging & Social Policy, 29*(1), 35–50.

Siibak, A., & Tamme, V. (2013). 'Who introduced granny to Facebook?': An exploration of everyday family interactions in web-based communication environments. *Northern Lights: Film & Media Studies Yearbook, 11*(1), 71–89.

Statistics Finland. (2017). *Suomen virallinen tilasto (SVT): Väestön tieto- ja viestintätekniikan käyttö* Retrieved from http://www.stat.fi/til/sutivi/index.html.

Taipale, S. & Farinosi, M. (2018). The big meaning of small messages: The use of WhatsApp in intergenerational family communication. In J. Zhou & G. Salvendy (Eds.): *Human aspects of IT for the aged population 2018, lecture notes in computer science* (pp. 532–546). Cham: Springer.

van Deursen, A. J., & Helsper, E. J. (2015). A nuanced understanding of internet use and non-use among the elderly. *European Journal of Communication, 30*(2), 171–187.

van Deursen, A. J., van Dijk, J. A., & Peters, O. (2012). Proposing a survey instrument for measuring operational, formal, information, and strategic internet skills. *International Journal of Human-Computer Interaction, 28*(12), 827–837.

van Deursen, A., & van Dijk, J. (2011). Internet skills and the digital divide. *New Media & Society, 13*(6), 893–911.

van Deursen, A. J., & van Dijk, J. A. (2015). Internet skill levels increase, but gaps widen: A longitudinal cross-sectional analysis (2010–2013) among the dutch population. *Information, Communication & Society, 18*(7), 782–797.

Van Dijk, J. (2012). *The network society.* London: Sage.

Venkatesh, A., Dunkle, D., & Wortman, A. (2011). Family life, children and the feminization of computing. In R. Harper (Ed.), *The connected home: The future of domestic life* (pp. 59–76). Dordrecht: Springer.

Wartella, E., Rideout, V., Lauricella, A. R., & Connell, S. (2013). Parenting in the age of digital technology. *Report for the center on media and Human development school of communication Northwestern University.* Eveston, IL: Northwestern University. Retrieved from https://cmhd.northwestern.edu/wp-content/uploads/2015/06/ParentingAgeDigitalTechnology.REVISED.FINAL_.2014.pdf.

Watt, D., & White, J. M. (1999). Computers and the family life: A family development perspective. *Journal of Comparative Family Studies, 30*(1), 1–15.

Zhong, Z. J. (2011). From access to usage: The divide of self-reported digital skills among adolescents. *Computers & Education, 56*(3), 736–746.

Wilska, T.-A. & Kuoppamäki, S. (2018). Necessities to all? The role of ICTs in the everyday life of the middle-aged and elderly between 1999 and 2014. In S. Taipale, T.-A. Wilska, & C. Gilleard (Eds.), *Digital technologies and generational identity: ICT usage across the life course* (pp. 149–166). London & New York, NY: Routledge.

Chapter 4
Beyond Social and Family Generations

Abstract Here the theoretical foundations on which the arguments in the book are built are developed. The chapter begins by introducing the concept of generation as both a cohort-based and a family-based construction. A discussion then follows of how various forms of intergenerational solidarity and conflict shape the relationships between family generations. Particular attention is paid to the need for an approach that goes beyond any strict generational division and is more sensitive to the ways in which individual lives are interconnected through the use of digital technologies. To assist in this task, a post-Mannheimian approach to generational identity is outlined.

Keywords Cohort · Family generation · Generation · Intergenerational solidary · Life course · Linked lives

Thus far, this book has discussed digital family and its social relationships as something actively 'done' and shaped through, and in interaction with, digital media and communication technologies. In addition to other consequences already sketched out above, such an everyday-life approach to the use of new technologies in the family context has implications also from the point of view of sociological theories of family generations. In this chapter, the dynamics of intergenerational relationships in digital families are considered in the light of a post-Mannheimian approach to generations as outlined in Taipale, Wilska and Gilleard's *Digital Technologies and Generational Identity: ICT Usage Across the Life Course* (2018). The basic components of this new theoretical framework are identified, suggesting that 'generationing'—the process whereby the social identity of a generation is produced—is by its nature nonlinear and intertwines with human life stages and important life transition points that may, in turn, activate or inactivate the use of certain technological tools and application in digital families. Before doing that, however, it is imperative to understand the strengths and limitations of the established generational concepts and their related approaches.

© Springer Nature Switzerland AG 2019   41
S. Taipale, *Intergenerational Connections in Digital Families*,
https://doi.org/10.1007/978-3-030-11947-8_4

Social Generations

To many, the concept of 'generation' is closely associated with Karl Mannheim's seminal work on the theory of generations (Mannheim, 1952). Mannheim's fundamental observation was that there was a gap between the values young people learnt from their parents and the reality that they themselves lived through and experienced. In examining the kind of generational differences making up this gap, Mannheim came up with his well-known distinction between generation as *location* and as *actuality*. A generation's location in time is naturally defined by its members' year of birth. Being born and living their formative years of youth during the same period of time enables individuals, at least potentially, to acquire a common understanding of who they are. For Mannheim, namely, to belong to a certain generation is also to occupy a *social* location, as that location may shape a person's self-consciousness the same way a class position or culture can. Thus, when a group of individuals of similar ages collectively lives through certain historical key events and experiences them in the same way, it can develop a generational consciousness, implying that its generational potential is actualized.

In Manheim's thinking, youth is then the main formative period when a collective generational consciousness is or can be produced. In later years of adolescence, young people process their surroundings with their peers and for themselves, contrasting their observations with those of their parents. This process of *generationing* may then result in a distinct generational consciousness. The shared social location can translate into new and creative reactions and adaptive strategies that help a generation to recognize its own position in contemporary society (Edmunds & Turner, 2002a; Elder, 1974). Sometimes, tangible changes in the political and social climate can trigger even quite fierce intergenerational conflicts between one generation and its parental generations (see, e.g. Edmunds & Turner, 2002b). While major events like a student uprising, civil rights protests or the conquest of space in the 1960–70s no doubt heavily contributed to the generational consciousness of the current post-war generation—the so-called baby boomers—it is less clear to what extents, for instance, new technological innovations such as personal computers and smartphones have influenced a 'we' sense for younger age cohorts who grew up experiencing the transformative power of digital technologies first hand in their youth.

Many sociologists have attempted to categorize successive generations based on both historical analysis and people's own perceptions concerning their generational belongingness (e.g. Roos, 1987; Strauss & Howe, 1991). In the latter regard, empirical evidence from, for instance, Finland suggests that older people more readily than younger people identify themselves as belonging to the same generation with their same-age peers (such as the Baby Boomers; see, e.g. Sarpila, 2012). There are at least two explanations for why this should be so.

First of all, it takes time to build a shared understanding of who 'we' are. Older generations have an advantage here in that more time has passed since their formative years (Bolin, 2016). The more time passes by, the more one has a chance to commemorate the key events from those years and thus inculcate in one of their significance.

Mass media, popular culture and historiography recurrently bring back into public discussion major historical events and phenomena that have shaped generational consciousness (e.g. the two World Wars, the fall of the Berlin Wall, the first Moon landing, major pop culture events like Woodstock, the Beatles, the Rolling Stones, etc.), promoting processes of commemoration (see, e.g. Bennet, 2009; Bolin, 2016). Drawing upon survey data from Finland, Sarpila (2012) has, however, shown how young people, as they age, might end up also reconsidering their generational identity, beginning to identify with different generational labels. As she found, in 1999, 31% of the queried Finns aged 20–29 felt themselves belonging to the 'IT generation', while 10 years later, in 2009, no more than 11% of those in the same age group felt the same. In the latter year, it was, interestingly, again the (then) 20–29 year olds who thought of themselves as the 'IT generation'. This finding is in line with the stereotypical notion that information technologies belong to youth. When people age and leave their youth behind them, they unavoidably come to face situations where they must reconsider what is or is not unique and special about just them as an IT generation, vis-à-vis the subsequent generations that are similarly, or even more, immersed in the digital world.

A second reason for why older people may more readily than young people identify generationally with their same-age peers has to do with the whole host of new, thematically overlapping generational labels that have emerged in the last few decades. Among these are, for just a few examples, denominations such as 'Net Generation' (Tapscott, 1998), 'Digital Generation' (Buckingham, 2006) and 'Digital Natives' (Prensky, 2001). This great diversity of available designations that all cover temporally overlapping phenomena may, namely, complicate the formulation of a solid, shared generational consciousness among young people. This circumstance, combined with the individualized life trajectories, personal networks and personalized consumption of media that characterize our time, stands in the way of widely shared, overarching key experiences that might then stamp entire age cohorts, the same way that exposure to key mass media events and spectacles functioned for us in the past. As concerns their basis in scholarship, moreover, the criticism here has also been that many of the designations or labels resorted to are not based on any systematic research, that they are overly narrow in their scope, and that, as a result, they reflect commercial interests rather more than any even potentially shared generational identity, as in the case of, say, the 'MTV Generation' or the 'Nintendo Generation' (Guzdial & Soloway, 2002).

The majority of such technology-related, or technology-specific, generational labels are not, and cannot be, defined as related to any successive time periods since they clearly coexist in time and are thus hard for an entire age cohort to identify with. Indeed, as Burnett (2010) has noted about the temporal aspect of generations, they are a movable feast. While, traditionally, a generation has been considered to cover a time period of approximately 15–25 years (e.g. the 'Lost Generations' of 1883–1900, the 'Baby Boomers' of the mid-1940s to 1950s/early 1960s), the more recent generational categories only refer to a period of 10–15 years (e.g. the 'Generation Y', from the early 1980s to the mid-1990s or the 'Generation Z', from the mid-1990s to mid-2000s). The most recent trend in naming generations by marking them with consecutive alphabets only, as in Generation Z, Y and Z, makes it obvious how generations as currently labelled cannot provide a good basis for generational identification: alphabets as such do not tell anything about who 'we' are as a generation.

In order to better understand the ability of young age cohorts to collectively identify themselves with a particular generation, we might do well to go back and retrieve another term Mannheim coined: *generation unit*. Generational units are smaller groups, fragments of an actual generation that develop different reactions to the same cultural and historical events (Mannheim, 1952). Proceeding from this conceptualization, it might then be possible, for instance, that the 'IT Generation' and the 'Digital Generation' are actually smaller units of one and the same generation, describing two subgroups of it that simply experience different aspects of a digital society as significant to them. They consist of people for whom the same technology is experienced as 'key' albeit from different angles, serving as it might different purposes in their lives. For instance, to some in this overall generation it may be the common utilization of social media platforms that form the basis of their 'we' sense, while for others in it is digital gaming that provides a sense of unity with one's coevals. This kind of internal fragmentation of larger cohorts seems to be one of the distinguishing features of our technology-rich and individualized contemporary cultures.

Family Generations

In addition to its Mannheimian definition as a cohort with a social-historical meaning, the concept of generation also has another distinct sense, involving kinship. In its classical sense, the notion of family generations underscores the meaning of blood relationships and marriage. Within the context of the family, a generational position is defined by a system of lineage and descent (see, e.g. Burnett, 2010). Traditionally, families have been seen as established by the marriage of two spouses, whose descendants then form the next familial generation. Given the actual diversity of families in terms of their shapes and forms, however, it seems obvious that this definition is insufficient and outdated, in Europe as elsewhere in other parts of the world. More and more often today, families are set up also between non-married partners—of either the same or different sex—and their children. New intergenerational family relationships are also created through series of divorces and remarriages, conjoining people from different family backgrounds in most varied manners.

Within families, it is the individual persons' relationships to elder family members, one's own siblings and children, and the possible partner(s) that are formative of their generational identity. A kinship system ensures that each new family member is immediately located in a network of family relationships, a family tree that fosters relatedness and belonging within the family. Concepts such as 'brother' and 'sister' underscore the closeness of the relationship in question, while attributive adjectives like 'great' and 'second' in family terms ('great-grandfather', 'second cousin') signal not only generational connection but also a relative distance compared to closest family members (Burnett, 2010, pp. 23–24). The pluralization of family forms has its own bearings on the family terminology, introducing new labels such as stepfather, stepmother, stepsister and stepbrother, which all imply both familial affinity and difference. Also, all such application of family terminology then contributes to the 'we' sense, helping to circumscribe who belongs to the immediate family and who to the extended one.

The new family relationships created by separations and remarriages make it more complicated to draw clear-cut generational lines between the members of many families. The father's new partner, the stepmother, might, for instance, on account of her age belong to the same age cohort as the family's grown-up children. Similarly, when the eldest sibling in a family with significant age differences between children becomes a mother or a father for the first time, the newborn baby might be of the same age as her youngest aunt or uncle. What complicates the notion of family cohorts even further, however, is that a person's generational position may be different in different family contexts. Being a member of two blended families such as when both of one's parents have established a new family, may mean that the same person is the youngest sibling in one family and the oldest in the other.

Against this backdrop, it seems clear that the pluralization of family forms, resulting from short-lived marriages and the destandardization and individualization of the human life course, forces us to rethink family generations in terms of other than just blood and kinship-based categories. Widmer's (2016) work on configurational families is helpful in this regard, underlining, instead of the traditional notion of family as a long-standing and coherent entity, the role of family ties that are cognitively and emotionally significant. As new family compositions emerge following divorces and remarriages, family ties become more variegated and diverse, challenging any notion of family generations as fixed categories. This does not, however, mean that family ties would become a matter of pure choice, or that generational distinctions and conflicts would become fully obsolete; it only implies that family generations have become more dynamic as categories subject to change and reconsideration across the entire human life course.

Given the significance of cognitive and emotional ties for the 'sensing' of the family, it is important to consider what kind of role new technologies may play in the maintenance of these ties. Studies have, for example, rather straightforwardly claimed family solidarity to have eroded particularly because family members are more individually networked via new media and communication technologies (e.g. Rainie & Wellman, 2012). Others, on the other hand, have proposed that, even though intergenerational family relationships on the whole are no longer governed by the normative ties of family solidarity the same way as in the past, digital technologies have introduced new means for enacting affectual and functional solidarity between those both near and afar Taipale, Petrovčič, & Dolničar, 2018). Dolničar and collaborators (2018), for instance, have shown how older people's engagement in assisted (or proxy) Internet use may to a large extent depend on the functional help and solidarity provided by younger family members, especially grandchildren. Irrespective of whether there may be more or less solidarity than before binding the members of contemporary extended families together, however, what seems clear is that at least some degree of solidarity, as well as a certain level of conflicts and ambivalence, remains characteristic of intergenerational relationships in all kinds of families and at all times (cf. Bengtson & Roberts, 1991; Lüscher et al., 2015). For this reason, theories of family solidarity, conflicts and ambivalence are vital for understanding the life of digital families and the relationships between family generations in them.

Perhaps the best-known work on intergenerational solidarity is Bengtson and Roberts (1991), published almost three decades ago already. The model developed in the book, drawing upon socio-psychological theories of sentiments and interaction as well as theories of social organization that highlight the importance of group norms and functional independence in behaviour, consists of six dialectical dimensions of solidarity. *Associational solidarity* alludes to the modes of interactions connecting family members across generations, ranging in their effect from integration to isolation. These modes include both spontaneous and ritual forms of communication with a varying degree of formality. *Affectual solidarity*, producing degrees of intimacy or distance, refers to the exchange of emotions and sentiments such as warmth, compassion and trust in intergenerational family relationships. The dimension of *functional solidarity*, influencing the degree of dependence versus autonomy, includes activities from financial assistance to immaterial help where the common denominator is the exchange of help. *Normative solidarity*, promoting different degrees of familism or individualism, refers to the endorsement of familial obligations, while *consensual solidarity* points to the degree of agreement within family with regard to beliefs, values or life orientations ranging from complete agreement to dissent. Finally, *structural solidarity*, providing opportunities or barriers through what is also known as the opportunity structure, refers to the availability of family members, which is dependent, for instance, on their physical proximity and health condition (Bengtson, Giarrusso, Mabry, & Silverstein, 2002; Bengtson & Roberts, 1991; Hammarström, 2005).

The original model of Bengtson and Roberts was grounded on the idea of 'idealistic' family relationships based on consensuality (Bengtson & Roberts, 1991; Bengtson, Rosenthal, & Burton, 1996). The model was, however, met with scepticism by, for instance, Lüschner and Pillemer (1998), who, deploying the concept of *intergenerational ambivalence*, pointed to the existence of contradictions between parents and their children that were not always resolvable. Confronted with this criticism, Bengtson and his collaborators (2002) went on to later modify their model so that it recognized conflicts and feelings of ambivalence both between and within family generations (e.g. Bengtson, Rosenthal, & Burton, 1996; Bengtson et al., 2002; Silverstein and Bengtson, 1997). The ambivalence noted, however, was seen to stem from structural and institutional (e.g. policy, cultural, economic) features intersecting with family life, thus still representing separate domains in fact.

While various forms of intergenerational solidarity and conflict thus shape the relationships between family generations, digital technologies and media consumption provide a new technological infrastructure for this mode of 'doing family' in extended and geographically distributed families. The ways in which 'our' generation and 'their' generation use digital technologies, and the kind of media contents children, parents and grandparents consume are tangible markers of generational differences in family life. Related to this, also diverging opinions about the 'right' and 'proper' ways of using the new technology are a typical source of generational conflicts and ambivalence in families.

Compared with social generations, family generations have one great advantage that facilitates their internal coherence and the level of agreement among them: the members of the same family have experienced many key events together, even if at different ages, supplying them with shared memories. Possibilities to recall where one was and with whom when something important happened serve as potentially important building blocks of social coherence in a family. While, earlier, the family photo album served as perhaps the most central tool enabling commemoration in families, today family members' Facebook timelines, Instagram accounts and smartphone photo galleries serve the same ends (see, e.g. Lohmeier & Böhling, 2017; van Dijck, 2008). Furthermore, given that, today, we often are part of more than one-family configuration, such personalizable and personalized online accounts have the additional benefit that they reflect any variations in the shared experience, not basing themselves on the assumption of one (homogenous) family the same way the family photo album most often does.

Individual Life Courses, Linked Lives

Due to the rapid pace of new digital technologies, building a solid generational consciousness around some single technology or application has become increasingly difficult. Generational experiences, such as of the arrival of radio and television that marked the landscape of domestic technology innovation for many years, have no longer been repeated in decades. The fast development of mobile communication technology is emblematic of this transformation. In the last 30 years, which equals just one-family generation, mobile phone networks have evolved from the first to their fifth generation. Over these years, the development and progress of mobile phones have come in both small steps and large strides, leading from simple feature phones to very complex multipurpose tools (see Taipale, Wilska, & Gilleard, 2018). While the arrival of the first personal mobile phones might have been a generational marker for young early adopters in the late 1990s, today the mere possession of a new smartphone model can hardly serve as the only, or sufficient, distinguishing factor for contemporary youth. The current generational markers in technology use more often have to do with differentiated contents, applications and ways of using personal communication technologies (Taipale, 2016).

In the family context, this constant influx of new communication tools and media equipment takes place according to family members' life stages. In this regard, previous research has identified many milestones in the human life course at which new technologies may become part of one's daily life. In many countries, parents typically buy their children their first mobile phones when they begin school or come of age. Similarly, the purchase of a laptop computer is often justified with reference to its potential educational benefits, and it is thus typically done for one's child when this reaches a certain educational level (Fortunati & Taipale, 2017). In adulthood, new devices and applications are adopted either owing to work-related duties or to keep up with one's children who use technologies increasingly independently while still needing some supervision (Ganito, 2018; Tammelin & Anttila, 2017). In later life, again, new digital technologies and applications may be acquired for recreational purposes when leisure time increases, to stay in touch with one's children and grandchildren, or to alleviate one's loneliness after retirement or loss of a partner (Ganito, 2018). In more advanced old age, monitoring and health technologies may be adopted for safety and security purposes or to prolong independent living at home.

Life-course studies, in general, have looked at the sequence of stages people live through as they grow older (e.g. Morgan & Kunkel, 2011). In social sciences, these stages centre on socially significant events that are formative for individual biographies such as changes in family roles and responsibilities (Shanahan & Macmillan, 2008). The different life stages are separated from one another by transitions, events such as entering and leaving school, gaining employment, getting married/divorced, moving abroad, retiring or widowing. Especially transitions specific to family life are of interest for family studies. Prior to the establishment of a new family, for instance, there is a courtship stage, followed by engagement and, finally, marriage or the beginning of cohabitation. Other major life-course markers in the family context are the birth of the first child, children's starting school, as well as their departure from home, along with a possible end of marriage/partnership or death of one's partner (Elder & Shanahan, 1997).

Unlike the rather fixed developmental life stages, the sequence of sociological life stages today is increasingly destandardized in its character. Instead of covering all people, life stages and transitions in our time involve constantly smaller and smaller parts of a population, or they are experienced at different ages and for varying durations (see, e.g. Brückner & Mayer, 2005). This destandardization of life-course patters has been explained by the transformation of our social and economical environments. After World War II, the project of rebuilding societies and stimulating economic growth favoured standardized life courses based on long-term or permanent employment contracts. Later, in the 1960s and 1970s, major demographic changes accompanied by cultural revolutions (e.g. the student movement, women's movement) paved a way for more heterogeneous family arrangements to emerge, altering the timing and sequencing of life-course stages. Latest by the 1990s, finally, economic uncertainties and high unemployment had begun to transform the structure of the labour markets, putting families under financial pressure and dissolving any remaining ideas of standard biographical trajectories of citizens (see, e.g. Zimmermann & Konietzka, 2017).

Although individual biographical trajectories have thereby become more diverse and variegated, one should nevertheless keep in mind that the lives of individual family members still today remain in many ways interlinked. The notion of linked lives, introduced by Elder (1994, 1998), implies that families are 'age-integrated': family members of varying ages, representing different birth cohorts, are joined together through their intermingling life trajectories. Such interconnectedness of lives is evident, for example, when one family member faces a major life transition. If a stay-at-home parent receives an employment offer from another city nearby, the decision to accept it will likely lead to the improvement of the family's financial situation. At the same time, however, it also forces changes in other family members' daily routines and care arrangements, in a most tangible manner. In order to succeed in surmounting such challenges, the interconnectedness of family members' lives requires a certain amount of family solidarity from everyone involved. Yet, it is also easy to see how family members' interdependence can also create *more* conflicts such as when the expectations of reciprocity or altruistic help provision within the family diverge (Blieszner, 2006). While some responsibilities in the families are passed on to the next generation(s) as people grow up, it seems likewise evident that certain family bonds based on kinship, affection and care are sustained throughout the life course: ageing parents keep caring also for their adult children, even when they no longer are responsible for the latter's daily lives, health and well-being.

Post-Mannheimian Generational Identity

A post-Mannheimian approach to generational identity builds upon the above-developed argument that, to understand the formation of generational identity today, attention to mere social generations (cohorts) is not enough; also the intertwinement of life courses with it, including the significance of their key transition points, and the effects of family generationing need to be acknowledged (Taipale, Wilska, & Gilleard, 2018). As people age, the relationships of dependence, interdependence and independence change within the family, which may render certain communication technologies and media tools unnecessary or irrelevant and create a new need for others. In what follows, the main features of such an approach to generational identity, attempting to update Mannheim's original conception of it, are briefly summarized.

To begin with, it is important that any work in this direction be premised on the observation that the technological identity of a generation does not emerge intrinsically with the passage of time (see, e.g. Buckingham, 2006). There is an *active process of 'doing'* behind the formation of every generation (McDaniel, 2007), involving continuous self-reflection and self-positioning in relation to other generations. While such efforts of 'doing' generation take place anywhere, at any time, the family is one of the main contexts for them. In families, similarities and differences between generations in technology adoption, technology use and the way individuals relate to technology occur naturally and are made visible. The family is also one of the few contexts in contemporary developed societies in which intergenerational interactions cannot be avoided. In it, generational differences in values, attitudes and digital technology usage patterns are constantly at issue and become thematized, leading to the boundary lines between generations to be defined and drawn.

Second, a post-Mannheimian approach to generational identity stresses the way a generational identity is *defined by the members and non-members of a given generation*. The characteristics of, and the criteria for, generational membership are defined by members sharing the same technology-related experiences who adopt and use the technologies in question in like ways. Although people's experiences may not be completely identical and their adoption and use patterns usually show some variation, their membership in the same cohort generation forms a major reference point for their own and others' generational self-positioning (Hepp, Berg, & Roitsch, 2017). However, the characteristics of a generation are also influenced by non-members—those who are either too young or too old to share the same generational experience. Perhaps the most tangible example of this pertains to the practices of labelling (other) generations. Quite often, adults (parents, but also researchers, marketing professional, media personalities) lapse to 'othering' new technologies and young people's practices of using technology, presenting these as unprecedented and transformational. In doing so, they reveal what seems to prevail in society even more broadly: an apparent discrepancy between adult perspectives and youth experiences. Consequently, many of the generational labels attached to young people as technology users ('Digital Natives', 'Nintendo Generation', etc.) tend to reflect adults' prejudices and stereotypes more than young people's own experiences or their own generational identity (see also Herring, 2008). Especially, in the family context, the unrealistic expectations of one generation regarding another one's technical skills and know-how may then lead to intergenerational disagreement and conflicts.

Third, the approach is suspicious of any static concepts of societal and family generations, viewing generationing as a *life-course-long process*, one in which certain periods, life transition points and single significant events are more formative than others. In this respect, the post-Mannheimian approach proposed here resembles Hepp, Berg, and Roitsch's (2017) processual conceptualization of media generations, which assumes the idea that generations evolve over time. This, however, does not mean that Mannheim's argument about youth as the key transformative period in generation building would somehow be discounted. Rather, it simply means that the years after youth are becoming increasingly more important as determinants of the technological identity of a generation. Due to the rapid digitalization of our contemporary societies, it is becoming increasingly difficult to age without engaging with new digital technology, services and applications. By extension, people's ability to adopt and independently use digital technologies in later life is increasingly more considered as a sign of their successful ageing. The extent and patterns of using digital technologies are more and more what determines one's generational position in relation to other generations, be these of the same age, younger or older.

A fourth and final reason for promoting a more dynamic approach to generational identity has to do with *family configurations*. As a consequence of divorces and remarriages, an individual's relative position in the family tree of generations may change. When belonging to several families at once, a person may be considered as a member of a digitally skilled generation in one family and as a digital latecomer in another. In a post-Mannheimian approach to generational identity in later life, major *life turning points* such as divorces, marriages, retirement, having one's first child or grandchild and other events of similar magnitude provide the formative events needed for generationing. They supply the need and reasons for the uptake or rejection of new technologies, and prompt specific practices and uses connected to these technologies. For instance, retirement may cause one to give up one's landline telephone and reduce the need for regular telephone calls, while the increased free time after it may motivate one to keep in touch with one's grandchildren via instant messaging or Skype, or engage in genealogical research on the Internet. Faced with such reconfigurations, a post-Mannheimian approach to generational identity can highlight the significance of life transition points and family life fractures, although not as factors for generational gaps, but as circumstances fostering 'for-the-family' and 'with-the-family' use of digital technologies (Taipale, Wilska & Gilleard, 2018) .

To conclude, a post-Mannheimian approach to the concept of generations helps us to understand the significance of life turning points after adolescence and in later life as formative elements of generational identity in the digital age. While the seeds of generational identity are planted while still young, each cohort generation has no choice but to over and over again reassess its technological self-understanding and reconsider its relative position vis-à-vis other generations, as new digital tools, applications and services are constantly being introduced that soon become prerequisites for a well-functioning independent life. For such a dynamic approach to generation studies to emerge, however, we first need to do away with stark generational oppositions (e.g. digital natives versus digital immigrants) along with any dualistic distinctions between right and wrong ways of using digital technologies not supported by empirical evidence (cf. Helsper & Eynon, 2010; Rosales & Fernández-Ardèvol, 2016). Only that way can we open up a perspective from which to rethink generational identity as malleable contract, one that can be adjusted, revised or refined throughout the entire course of life.

References

Bengtson, V., Giarrusso, R., Mabry, J. B., & Silverstein, M. (2002). Solidarity, conflict, and ambivalence: Complementary or competing perspectives on intergenerational relationships? *Journal of Marriage and Family, 64*(3), 568–576.

Bengtson, V., & Roberts, R. E. (1991). Intergenerational solidarity in aging families: An example of formal theory construction. *Journal of Marriage and Family, 53*(4), 856–870.

Bengtson, V., Rosenthal, C., & Burton, L. (1996). Paradoxes of families and aging. In R. H. Binstock & L. George (Eds.), *Handbook of aging and the social sciences* (pp. 253–282). New York, NY: Academic Press.

Bennett, A. (2009). "Heritage rock": Rock music, representation and heritage discourse. *Poetics, 37*(5–6), 474–489.

Blieszner, R. (2006). A lifetime of caring: Dimensions and dynamics in late-life close relationships. *Personal Relationships, 13*(1), 1–18.

Bolin, G. (2016). *Media generations: Experience, identity and mediatised social change.* London: Routledge.

Brückner, H., & Mayer, K. U. (2005). De-standardization of the life course: What it might mean? And if it means anything, whether it actually took place? *Advances in Life Course Research, 9,* 27–53.

Buckingham, D. (2006). Is there a digital generation? In D. Buckingham & R. Willett (Eds.), *Digital generations: Children, young people, and new media* (pp. 1–13). Mahwah, NJ: Lawrence Erlbaum Associates.

Burnett, J. (2010). *Generations: The time machine in theory and practice*. Farnham: Ashgate.

Dolničar, V., Grošelj, D., Hrast, M. F., Vehovar, V., & Petrovčič, A. (2018). The role of social support networks in proxy Internet use from the intergenerational solidarity perspective. *Telematics and Informatics, 35*(2), 305–317.

Edmunds, J., & Turner, B. (Eds.). (2002a). *Generational consciousness, narrative and politics*. Oxford: Rowman and Littlefield.

Edmunds, J., & Turner, B. (2002b). *Generations, culture and society*. Buckingham: Open University.

Elder, G. H., & Shanahan, M. (1997). The life course and human development. In R. M. Lerner (Ed.), *Handbook of child psychology: Theoretical models of human development* (pp. 665–715). New York, NY: Wiley.

Elder, G. H. (1994). Time, human aging and social change: Perspectives on the life course. *Social Psychology Quarterly, 57*(1), 4–15.

Elder, G. H. (1998). The life course as developmental theory. *Child Development, 69*(1), 1–12.

Elder, G. H. (1974). *Children of the great depression: Social change in life experience*. Chicago: Chicago University Press.

Fortunati, L., & Taipale, S. (2017). Mobilities and the network of personal technologies: Refining the understanding of mobility structure. *Telematics and Informatics, 34*(2), 560–568.

Ganito, C. (2018) Gendering the mobile phone: A life course approach. In S. Taipale, T.-A. Wilska, & C. Gilleard (Eds.), *Digital technologies and generational identity: ICT usage across the life course* (pp. 87–101). London & New York, NY: Routledge.

Guzdial, M., & Soloway, E. (2002). Teaching the Nintendo generation to program. *Communications of the ACM, 45*(4), 17–21.

Hammarström, G. (2005). The construct of intergenerational solidarity in a lineage perspective: A discussion on underlying theoretical assumptions. *Journal of Aging Studies, 19*(1), 33–51.

Helsper, E. J., & Eynon, R. (2010). Digital natives: where is the evidence? *British Educational Research Journal, 36*(3), 503–520.

Hepp, A., Berg, M., & Roitsch, C. (2017). A processual concept of media generation. *Nordicom Review, 38*(1), 109–122.

Herring, S. C. (2008). Questioning the generational divide: Technological exoticism and adult constructions of online youth identity. In D. Buckingham (Ed.), *Youth, identity, and digital media* (pp. 71–92)., The John D. and Catherine T. MacArthur Foundation Series on Digital Media and Learning Cambridge, MA: The MIT Press.

Lohmeier, C., & Böhling, R. (2017). Communicating family memory: Remembering in a changing media environment. *Communications, 42*(3), 277–292.

Lüscher K., Hoff, A., Lamura, G., Renzi, M., Sánchez, M., Viry, G., de Salles Oliveira, P. (2015). *Generations, intergenerational relationships, generational policy. A multilingual compendium*. Retrieved from http://www.kurtluescher.de/downloads/Luescher-Kompendium_7sprachig-komplett_online_15-10-2015.pdf.

Lüscher, K., & Pillemer, K. (1998). Intergenerational ambivalence: A new approach to the study of parent-child relations in later life. *Journal of Marriage and Family, 60*(2), 413–425.

Mannheim, K. (1952). Essay on the Problem of Generations. In P. Kecskemeti (Ed.), *Essays on the sociology of knowledge by Karl Mannheim* (pp. 276–320). New York, NY: Routledge & Kegan Paul.

McDaniel, S. (2007) *Why generation(s) matter(s) to policy*. Working paper 2017-11-22. Institute of Public & International Affairs. Salt Lake City: University of Utah.

Morgan, L. A., & Kunkel, S. (2011). *Aging, society and life course*. New York, NY: Springer.

Prensky, M. (2001). Digital natives, digital immigrants. *On the Horizon, 9*(5), 1–6.

Rainie, L., & Wellman, B. (2012). *Networked: The new social operating system*. Cambridge, MA: MIT Press.

Roos, J. P. (1987). *Suomalainen elämä. Tutkimus tavallisten suomalaisten elämäkerroista*. Helsinki: SKS.

Rosales, A., & Fernández-Ardèvol, M. (2016). Beyond WhatsApp: older people and smartphones. *Revista Română de Comunicare şi Relaţii Publice, 18*(1), 27–47.

Sarpila, O. (2012). Minun sukupolveni, sinun sukupolvesi. *Hyvinvointikatsaus: sukupolvien väliset suhteet* (pp. 14–18). Statistics Finland: Helsinki.

Shanahan, M. J., & MacMillan, R. (2008). *Biography and the sociological imagination*. New York, NY: W.W. Norton.

Silverstein, M., & Bengtson, V. L. (1997). Intergenerational solidarity and the structure of adult child–parent relationships in American families. *American Journal of Sociology, 103*(2), 429–460.

Strauss, W., & Howe, N. (1991). *Generations*. New York, NY: Harper Perennial.

Taipale, S. (2016). Synchronicity matters: Defining the characteristics of digital generations. *Information, Communication & Society, 19*(1), 80–94.

Taipale, S., Petrovčič, A., & Dolničar, V. (2018). Intergenerational solidarity and ICT usage: Empirical insights from Finnish and Slovenian families. In S. Taipale, T.-A. Wilska, & C. Gilleard (Eds.), *Digital technologies and generational identity: ICT usage across the life course* (pp. 68–86). London & New York, NY: Routledge.

Taipale, S., Wilska, T.-A., & Gilleard, C. (Eds.). (2018). *Digital technologies and generational identity: ICT usage across the life course*. London & New York, NY: Routledge.

Tammelin, M., & Anttila, T. (2017). Mobile life of middle aged employees: Fragmented time and softer schedules. In S. Taipale, T.-A. Wilska, & C. Gilleard (Eds.), *Digital technologies and generational Identity: ICT usage across the life course* (pp. 55–68). London & New York, NY: Routledge.

Tapscott, D. (1998). *Growing up digital: The rise of the Net Generation*. New York, NY: McGraw Hill.

van Dijck, J. (2008). Digital photography: Communication, identity, memory. *Visual Communication, 7*(1), 57–76.

Widmer, E. D. (2016). *Family configurations: A structural approach to family diversity*. Abingdon, Oxon, NY: Routledge.

Zimmermann, O., & Konietzka, D. (2017). Social disparities in destandardization—Changing family life course patterns in seven European countries. *European Sociological Review, 34*(1), 64–78.

Part II

Roles, Responsibilities and Practices

In this second part of the book, the analytical focus is moved from concepts and theories to new roles, responsibilities and practices making themselves manifest in the everyday life of digital families. At the centre of my examination is the observation made in this study that although families in the three countries included in it have, on average, become digitally better equipped and more skilled, their digitalization has not proceeded simultaneously, at the same pace, and along the same paths. Especially when looking at the extent to and ways in which older family members utilize new media and communication technologies, many differences surface. However, country differences are also apparent in the way families appropriate digital technologies and how the use of new technology is seen as influencing shaping family roles, household tasks and responsibilities, and intra-family communication practices.

Of the three countries in this study, it was most common to have a family with at least three generations using basic or more advanced digital communication technologies in Finland. Ordinarily, these included key informant, his or her parents and grandparents. Even then, however, it was typical that grandparents used a much narrower range of digital technologies and a more limited number of functions in their mobile phones and other devices than the rest of the family (e.g. the grandparents in the families of Maria and Simon). On the average, they were also less attached to their personal digital devices than the younger family members, feeling, for example, that the mobile phone should not always be carried along (e.g. the families of Jenny in Finland, Alexander and Anton in Slovenia). Some older respondents were also very content with their basic feature phones and had found, for instance, a desktop or laptop computer to be more suitable for their needs than a smartphone. This was the case in the digital family of the Finnish Simon, whose grandmother used Skype to keep in touch with him. As Simon explained, 'None of my grandparents owns a smartphone. With my Skype-using grandmother (aged 70) our Internet-based communication is pretty variable, because she doesn't go on the Internet on a daily basis'. Another Finnish key informant, Benjamin (aged 29), told that 'My grandma calls her daughter (aged 60+) every day on Skype, to exchange news. Those calls are part of their daily social interactions'.

Also in Italy and Slovenia, many grandparents had started using a mobile phone, although landline equipment was still more widely used in these two countries than in Finland, where only a couple of the interviewed persons had a fixed-line telephone at home. In that country, the number of fixed-line telephone subscriptions has dramatically dropped over the last decades. In 2005, there were almost 1.5 million household subscriptions, while in 2017 only 151,000 were left (Finnish Communications Regulatory Authority, 2018). The Finnish interviewees who reported having and using a landline telephone were typically also mobile phones users. This was in stark contrast to Italy, where one of the interviewees, Matteo, described his typical family situation as follows: 'My grandmother does not use any ICTs. The most technologically advanced device that she's able to use is the landline phone'. In Slovenia, Alexander's family was rather similar: 'I sometimes call my grandpa [aged 70] and my grandma [aged 64], who are more than 40 years my senior and live together, on their landline phone, which they still use'.

With the exception of older people's stronger attachment to landline phones, the interviewed Italian and Slovenian families were relatively well equipped with new digital communication technologies, such as mobile phones, desktop computers, laptop computers and tablet computers. The Slovenian Sebastjan's (aged 26) description of his family's technological arsenal serves as a good illustration:

> My mom [aged 47], my dad [aged 50], and my sister [aged 21] consider themselves very ICT savvy. They all use smartphones, PCs, and laptops on a daily basis, so they feel comfortable using them. In addition to that, my father regularly uses a tablet and he considers himself the most adept in the family at operating it.

What was already noted above, that the uptake speed of new digital technologies, just as the extent to and ways in which these technologies were utilized, varied considerably between the three countries studied, applied also to the situation within each country. Furthermore, there were also families in them in which new technology, according to the key informants, did not play any notable role at all. In these families, people were either not eager in general to 'go for new gadgets' or they felt new communication technologies to offer very little that was useful to their family, as was often the case in Slovenia where families continue to live in close proximity to one another.The following chapter takes a closer look at the family roles in digital families, simultaneously drawing upon and attempting update for the present day the concept of the *warm expert* as introduced by Bakardjieva (2005) already more than a decade ago. Of particular interest in it is how digital families negotiate and assign specific roles to different family members, and which family members they rely on when solving technology-related problems. The specific ways warm experts co-use new technologies with those in need of assistance are described. In the subsequent chapter (Chap. 6), a more detailed analysis of new household task and responsibilities following the digitalization of the home is then presented. Many of these tasks and responsibilities fall, albeit not exclusively, on the shoulders of warm experts. Chapter 7 narrows the discussion to one specific communication practice widespread in digital families today. The chapter explores the use of WhatsApp freeware use in digital families, considering in particular the importance of short text and voice messaging for the development and maintenance of a sense of togetherness in geographically distributed digital families. Part II ends with Chap. 8 that considers whether and how changes in the maintenance of digital home and familial relationships might be linked to the ways in which intergenerational family solidarity is expressed in extended families.

Reference

Bakardjieva, M. (2005). *Internet society: The internet in everyday life*. London: Sage.

Chapter 5
Warm Experts 2.0

Abstract This chapter focuses on family roles in digital families, drawing upon, and updating for the present day, the concept of the warm expert. First, the impact of information and communication technologies on family roles is investigated, based on qualitative research material collected from Finland, Italy and Slovenia in 2014 and 2015. After that the analysis looks at how family roles and responsibilities can change over the human life course. Three types of warm experts are identified, with their characteristics described and discussed. Lastly, the argument is made that intimately knowing the other family members is an essential quality of those acting in the role of warm experts, and that while acting in the role of an warm expertise is often demanding, it can also be rewarding to not just those benefiting from it, but also those in it.

Keywords Family relationships · Family roles · Information and communication technology · Life course · Proxy user · Technology co-use · Warm expert

The concept of the warm expert has continued to attract many new media and communication researchers. Introduced by Bakardjieva (2005), it was intended to help investigate the first wave of ordinary technologies in the early 2000s that allowed people to access the Internet from their homes. Warm experts, for Bakardjieva, were people with relatively advanced skills and knowledge about new technology who were readily available to assist novice technology users taking only their first steps in using digital technologies.

In contrast to outside professional helpers—'cold experts'—the warm experts Bakardjieva (2005) described share their daily lifeworld with people needing their help, and are thus readily at hand for them to demonstrate, drawing upon their own experience, the advantages of being digitally connected. In her study, such everyday help and support by these experts were typically favoured over professionalized forms of assistance, including those provided through telephone helplines and computer service shops. The larger the knowledge gap between the helper and the helped, however, the higher the threshold for asking assistance would appear to be (Barnard, Bradley, Hodgson, & Lloyd, 2013). This much seemed evident also from the account provided by the Finnish key informant Lucas (aged 38). As he reported, his mother-

© Springer Nature Switzerland AG 2019
S. Taipale, *Intergenerational Connections in Digital Families*,
https://doi.org/10.1007/978-3-030-11947-8_5

in-law found it much more convenient for her to ask for help from her own children than, for instance, from a relative who worked as an IT professional:

> When it comes to [mother-in-law's, aged 62] mobile phone use, it's been one of her children who's been helping her, for example by showing her how you save phone numbers to the phone memory and how to turn off auto correction for text messages. One of her male relatives is an IT professional and so could certainly help with a lot of things, but she finds his IT advice to often be pretty difficult for her to understand, so she rather turns to her own children in these matters.

After Bakardjieva's seminal study, the landscape of new media technologies has transformed quite radically, however. The formerly predominant stationary equipment has been replaced by small-sized and mobile personal devices that are today everywhere. In families, the primary function of warm experts is no longer to convince other family members about the usefulness of the new technologies, but to help others to update their devices, keep up with technological developments and manage software contents and applications. Such *digital housekeeping* has come to form an essential part of families' everyday life and a prerequisite for their smooth functioning, as will be noted in more detail in Chap. 6.

Given the changing domestic technology landscape, the role of warm experts must be revisited. The rest of this chapter is dedicated to an examination of how the role of the warm expert is assigned, adopted and performed in digital families today, a decade and a half after Bakardjieva's original formulation of the term. As I will show, the role is still typically assigned to one of the younger family members, who in turn appreciate the recognition of their usefulness as one of merit. However, among these younger family members the role of the warm expert also entails the presence of contradictory feelings. On the one hand, in digital families, there is an expectation that all family members should continuously learn and develop new personal skills, and hence be able to sort out at least some of the technical problems they face in their IT use on their own. On the other hand, younger family members, too, are aware of the limited nature of the digital skills they personally have, confined as those typically are to certain technologies, applications and operating systems only. In consequence, when warm experts are unable to provide the assistance they would like when called upon to help, feelings of inadequacy arise.

Family Roles: What Has Changed?

The key informants for this study in Finland, Slovenia and Italy were instructed to investigate, among other things, how ICTs shaped or had shaped family members' roles within their own families. In doing so, they were to look at all kinds of digital communication tools and applications used today for the purposes of staying in touch and communicating with other family members—mobile telephony, email, Facebook, Twitter, WhatsApp, and Instagram, and the like. Although the informants' views regarding this seemed extremely diverse at first glance, with much variance in the understandings about the matter also within families, two main themes could nevertheless be gleaned from their reports.

First of all, there was a double suggestion that ICTs had not changed family roles at all, or had done so only very little, but that they had nonetheless transformed the way families communicated within themselves. This was an understanding put forth by key informants in all three of the countries studied, with some of them appending a view that family roles and relationships are, at least to a certain degree, structured by the human life course and thus given, for which reason new technology as such could not change anything essential in them. As the Slovenian key informant Natalija (aged 30) put it, 'Family roles are assigned at birth, which means that parents will always be parents and children will always be children, and the roles don't change when we look at them through the prism of the ICT use.'

Another informant from Slovenia, Sandra (aged 25), provided an account of how family roles change as time passes, claiming, however, that this had nothing to do with new technologies. According to her, it was part of the normal life course that parents act as their children's teachers, but that when children grow up, they are given, and themselves claim, more freedom:

> The use of ICT technologies has not changed any roles in our family. The same was said also by my uncle [aged 46] and my grandfather [aged 82] …. Regarding who buys the mobile phones, when my sister and I were younger, it was our parents who decided for us. Now we choose ourselves, discussing it amongst ourselves without consulting our parents, since our parents are not so up-to-date in these matters.

There were also families who went on to provide reasons for why they thought technology to not have changed family dynamics in them. In Finland, for instance, the key informant Paula (aged 26), explained that, in her family, 'communication technology was not seen to have any profound effect in family roles...probably because some sort of dialogue and open sharing of opinions has always been a feature of our family life'. Somewhat along the same lines, Ella (aged 24), another key informant in Finland, stated that '[t]echnological devices don't have an impact on those roles; they only facilitate communication'. Italy was no different from Finland in this regard. There the key informant Antonio (aged 30), for example, reported everyone apart from his grandfathers to think that ICTs did not have a significant influence, 'as long as they don't completely replace all immediate physical relationships'. Similar views presenting relatively unaffected family roles were put forth also by Sabrina (aged 23), another key informant in Italy:

> While the use of ICTs has not at all changed the roles of parents and children in our family, ICTs have nevertheless changed the way we communicate: WhatsApp, for example, tends to bring a certain playfulness to both the conversations topics and the tone of our messages, so, I'd guess, it makes our dialogue kind of friendlier in nature than how it usually is in exchanges between parents and children.

There was, however, another view that held that digital technologies indeed had changed the family roles, but only in the realm of technology use. In many of their accounts, key informants sought to maintain that the impact of digital technologies on intergenerational relations had remained limited at most, and anyway only concerned some narrow areas of life. For example, the Italian key informant Martina (aged 21) proposed that ICT had made it possible that 'the younger generations are àble to control a much broader area [of their lives]' than before, but no more than that. Interestingly, even when parents and grandparents seemed to think that child–parent relationships had changed, younger family members were reluctant to admit to such changes in family roles. The point was eloquently put by Carla (aged 23) from Finland:

> In the interviews with my parents, this thing came up that the 'the egg is smarter than the chicken'.... The younger people in the family think that technology doesn't influence or change anything about the roles of the family members. Rather, it brings family members more to the same level, so to speak. One can, let's say, teach some skills to the others, and, on the other hand, people can together think what the good and bad aspects of devices and software might be.

In some families, there was a firmer consensus about the impact of technology on family roles. Claudia (aged 21) from Italy, for instance, reported a view in her family that 'teaching the use of ICTs does bring new roles for family members on a general level, but only in relation to communication and the purchase of new mobile devices or PCs for using the applications'. Elaborating on the same observation, the Slovenian Tina (age 25) put forth that '[r]egarding the impact of ICTs on the roles of the individual members of our family, we have noticed a bit of a reversal in them. When it comes to technology, the children have definitely taken over the main roles in the family.'

A closer look reveals that the changes identified in family roles were, in particular, linked to how the educational relationships within the family looked like and, in some cases, how hardware purchase decisions were made. The Finnish key informant Isabella (aged 22) was very outspoken on this point, explaining that it was no longer the oldest person in the family who had the final word:

> Technological skills greatly affect the roles within the family, at least in situations where technology-related problems are being discussed. In them, the tasks that previously belonged to the head of the family are handed over to the person who has most knowledge, and even the oldest member of the family, which is our grandfather, is no longer listened to. Usually the one who is asked to take over is me. Normally in my family, it's the father who makes all the big decisions, then the mother, and the children are only listened to after all the others have had their say. But of course, it's become a bit different also for other reasons since the children have become adults.

In Slovenia, the key informant Sandra (aged 25) explained that '[W]hen it comes to choosing and actually buying, say, a new mobile phone, our parents turn to us and we tell them what we think about the quality and usefulness of the new phone they're maybe having in mind.' Overall, the Slovenian informants were more careful in their assessments regarding possible changes in their family roles, compared to their Finnish and Italian counterparts. As one of them, Mia (aged 25), for instance, stated, 'ICTs change family roles only in the sense that parents begin to more often have certain kinds of questions for us, or things that they want to learn from us, because we are young.'

Changes Across the Life Course

Although perceptions about the impact of digital technology on family roles varied from country to country, there was nonetheless a fairly uniform view that family roles, in general, change over the human life course. This was seen to apply to every facet of family life, including also the appropriation and use of digital technologies. When family members reach a certain life stage or pass through certain key points in their lives, the role of the warm expert is usually passed from one generation to another. As in life more in general, both young children and old persons are dependent on the help of others also in their technology use, but young people gradually grow more independent as they gain more knowledge.

First, however, when children are small, parents serve as warm experts for their minors. The key informants in this study who had small children typically already anticipated in their accounts how and when they would begin guiding their youngsters into the world of mobile communications and the Internet. This was the case, for instance, with the Slovenian Katarina (aged 26), who had a 1-year-old son whom she expected to soon start to become interested in her smartphone:

> I'm planning to teach him how to use the Internet safely, and also how to use ICTs like the mobile phone, tablet computers, and PCs. When he's a bit older, of course, maybe four or five, I'm going to start teaching him how to play educational games, maybe also some games developed by the company that my partner and my father work for.

Another Slovenian key informant, Tina (aged 25), noted that her female cousin was already teaching her 3-year-old daughter to play games and watch cartoons on YouTube. In Finland, Rita [aged 34], too, was already thinking ahead to what would happen when her daughter would be a little older. 'For starters', she explained, 'I and my husband are going to start slowly teaching her how to use the phone and text; she'll get her own phone when she starts school.' A little later in her report, Rita shifted her attention further ahead: 'I'm thinking that as my daughter [aged 7] grows up, she will then later on teach me, like, how to use the latest programmes. Right now, she still needs a lot of adult guidance in the use of communication technologies.'

When reflecting on the changes in family roles and relationships, the key informants also looked back in time to how things used to be when they themselves were young. Rita from Finland made a note on this, writing that '[w]hen I was younger myself, I learnt from my parents, but today this is no longer how it goes.' Indeed, it was easy to see that teaching children the basics of current technology had always in the past been the parents' duty. This kind of intra-family help was regarded as something natural by the key informants, as the following quote from the report of the Slovenian Tina (aged 25) makes clear:

> It's somehow normal and it goes without saying that we now help and teach our parents, considering that they first raised us for so many years and taught us all the important things. Parents give advice to their kids when they're young, also in technology-related matters.

Also Italian and Finnish key informants explained that the way technology and technology use were taught in their families, and who was responsible for that was based on family members' current life stage. As the Italian Monica (aged 25) described it, 'Previously it was my father who would teach me how to use the technology, now I teach him.' In Finland, Teresa (aged 24) reported rather similarly that '[e]verybody in my family holds the view, however, that parents have to teach their children the basics of how to use computers and phones'. A little further on in her report, she specified that such 'teaching includes things like when the phone must be on or in silent mode, and how to answer the phone and take good care of your devices.' In other words, what she suggested was that the parents' role as teachers was limited to the very basics intended to help the children to be able to get started.

As children grow up somewhat, they become warm experts to one another. Often, the knowledge they have of new technology is shared with other siblings, along with the digital skills one has acquired. As Maria (aged 24) from Finland elaborated on this stage of the life course:

> We, the siblings, we got guidance from our parents very early on how to use the computer and telephones, but since those times we've learnt a lot on our own, experimenting by ourselves, and actually even more from those amongst us who were more advanced. My youngest sister said she got most help from the second youngest amongst us. On the other hand, sometimes that same youngest sister found some application or another that the rest of us, us older sisters, had never even heard about before.

Later, when parents and grandparents grow older in the family, younger people's role as warm experts for them becomes 'officially' acknowledged. Their new role is not only owing to their more up-to-date knowledge of new devices and applications, it is also called for because of age-related cognitive and physical limitations experienced by older family members. A good example of this was provided by the Finnish key informant Emilia (aged 24):

> My grandpa's [aged 85] vision has gotten so much worse that he nowadays can barely read or write. And he has also begun to forget how different gadgets work, so he doesn't use anything anymore, other than his phone…. In the past, he told us, he used to use Skype a lot.

Older family members benefit from the support provided by the warm experts in the family in at least three different ways. To begin with, some older people receive continuous support from warm experts to help them sustain and sometimes even improve their digital skills in later life. The Finnish key informant Marika (aged 20), for instance, was able to note the following of the effect of such support on her grandmother (aged 75): 'she's not that bad [with her skills] anymore, she's improved; in fact she's constantly becoming better as a user, with help from us younger people'.

Second, other family members sometimes *co-use* digital technologies and applications with warm experts, being hence more dependent for their ability to put them to use on these members' physical presence. As Emma (aged 24) from Finland described one such situation: 'Sometimes when one of my grandmas comes visiting us, we use Skype together. That, they say, is very special and exotic for them, since they've never used any technological devices or made video calls before in their lives.' Another case, that of the Italian key informant Emilio (aged 30), involved an aunt who 'thanks to her children learnt to use the smartphone, the computer,

Facebook, and Skype', although she did not, for instance, have a personal Facebook account; instead, she used her sons' accounts.

Third, there are always some older family members who remain unable to use any of the new technology, whether on their own or with help of others. Such older people may then take advantage of so-called *proxy users* who use it for them or on their behalf (see, e.g. Dolničar, Grošelj, Hrast, Vehovar, & Petrovčič, 2018; Selwyn, Johnson, Nemorin, & Knight, 2016). In this study, this was the case, for instance, with the family of Emilia (aged 24) in Finland: in it, one grandfather's bills were paid online by a cousin of Emilia's. In Slovenia, too, such proxy use was described by the key informant Tia (aged 26), whose mother (aged 54) had expressed her desire that 'for the time being it's fine for her if any information that's only available on the Internet is accessed there by either my brother [aged 35] or me [Tia].' Also, the Italian key informant Matteo (aged 24) told that her grandmother relied completely on younger people for her technology use. According to Matteo, this grandmother, though:

> [w]ould like to learn how to use a computer, so that she could get a Facebook account and do Skype calls to her relatives in America, but then she keeps complaining that she still hasn't even figured out yet how to use her mobile phone. So then she gives up learning before even trying, and instead asks my cousins and me to do everything for her.

As the last two quotes reveal, when the challenges of learning a new technology are estimated to be significant, people are inclined to look for help. The support sought is then expected to help minimize the time and effort that learning the new technology would likely otherwise take (Barnard et al., 2013). In the context of extended digital families, the trusted persons in them, their warm experts, are then the first ones to resort to when the need for help or support in the use of digital media and communication technologies arises.

Three Types of Warm Experts

Who, more exactly, then acted as the warm experts in the geographically distributed and extended families in this study? Based on the key informant reports, the role of the warm expert was most often assigned to one or two persons in the family. In the family of the Slovenian Tia (aged 26), there was only one such person: 'As for teaching ICT use and introducing new ICTs in our family, my father (aged 67), my mother (aged 54), and my brother (aged 35) all agree that it's me who's to do it and ensure that everything is used properly.' (The same was true of the families of the Slovenian Mia, aged 22, and Aleksej, aged 25; the Italian Alice, aged 23; and the Finnish Sofia, aged 24, Lucas, aged 38, and Karin, aged 27).

The largest and most prominent category of warm experts were also in this study made up of younger family members who provided information to their parents and siblings on technical aspects and helped them with software issues. In the three countries in question, it was normally the key informants themselves, or one of their siblings, all aged between 20 and 35, who acted as warm experts in their families. In Slovenia, the key informant Erik (aged 25) described the role of such younger warm experts in his family as follows:

> My brother and I act as a source of information on all things related to ICT, answer questions like what does this thing do, what is this all about, how can I turn this machine on and off, and so on and so forth. We are especially sought after when the others need to upgrade to a new ICT device or newer software version. In short, we deal with the small problems encountered by the other family members who're not so interested in technology.

The key informants' parents, who were typically in their late middle age, were, however, actually quite often capable of installing basic digital devices such as televisions and laptops. As Veronika (aged 27) from Slovenia described her father (aged 52), '[he] is responsible for installing all devices and setting up their network connections. He also decides when and where to send a device for repair when that's needed.' When there were software issues or problems with some applications in the family, it was nevertheless the younger people who stepped in as trusted persons (see Software and Application Installations in Chap. 6).

Skipped-generation warm experts were the second category of warm experts who could be identified in the key informant reports. The term refers to the help provided by grandchildren to their grandparents without parents' involvement in the interaction. As such it, in fact, describes a situation opposite to that alluded to by its root adjective 'skipped generation', which is used for situations where grandparents raise their grandchildren in the absence of parents. Skipped-generation help in technology use was in this study, especially, common in Slovenian families, in which multiple generations frequently lived on the same property and even in the same house, or otherwise physically proximate to one another. Compared to their Italian and Finnish counterparts, grandchildren in Slovenian families were thus to a far larger extent available to their grandparents and their needs. Such close intergenerational relationships were exemplified by the Slovenian Mia (aged 25), who wrote that '[m]y grandfather said that he prefers to ask his grandchildren since we know where to look when something is wrong with his mobile phone and we know what he is trying to say.' The similarly Slovenian Klara (aged 28) told about how 'my younger sister [aged 24] and I taught our grandmother [aged 80] how to use a mobile phone designed for the elderly'. A third Slovenian informant, Julija (aged 25), described more at length what kind of help she and her cousin provided for the oldest members of her family:

> When I visit my grandparents, I often show them how to use certain ICTs. Recently, I taught my grandfather [aged 70] how to save images from his digital camera to his laptop computer, and how to use a programme for viewing photos electronically. It's often the case that I also advise my mom on similar issues. In addition, my younger cousin [aged 18] often teaches my grandfather and grandmother to use ICTs.

In the reports by the Italian and Finnish key informants, there were only few examples of skipped-generation warm experts and help. As regards Italy, Silvia (aged 25) stated that her great-uncle often had problems with his computer and called her up to have her help solve them. In Finland, the relationships between grandchildren and grandparents were even more sporadic and distant. An exception in this regard was the family of the key informant Isabella (aged 22), who described her grandfather as very receptive to new technology, which made him to request her help from time to time. In her report, Isabella noted how '[l]ast time we learned how to use email, which has been my regular means of communication for a long time already. My grandfather, however, had never used it before, so he specifically asked me to help him with it'.

The third category of warm experts in my material consisted of older family members, who were either the key informants' parents or grandparents serving as warm experts for their age-mates. While peer support and learning are commonly associated with young people, also older age-mates provided support in digital technology use in the three countries studied. Moreover, it was both men and women who supplied this help to their spouses. The Slovenian key informant Angela (aged 27), for example, described such warm expert support between her parents. Starting with the characterization 'my mother comes to the rescue of my father when he needs someone with ICT skills', she then concluded, 'it's enough if one of the parents is skilful in the use of ICT tools'. In the family of the Finnish key informant Rita (aged 34), it was Rita's father who served in this role vis-à-vis Rita's mother: 'If she want to talk on Skype, she makes sure that my father is there to guide her through it.' For the Italian Monica (aged 25), too, it was 'my father who teaches my mother, even though she is younger,' demonstrating that was not just the person's age that mattered in this regard; more than that, it was personal interest, motivation and sense of attachment to certain technologies that determined who acted as the expert within the family. As the above Monica further explained, her mother felt emotionally quite attached to the family's landline phone, and so needed help from others when it came to using newer technology. Help-giving between approximately same-age persons was also common among the key informants' grandparents. The Slovenian Boris (aged 26) spoke about his grandmother (aged 79) as someone who 'thinks that she knows much more [about technology] than other people her age', leading to a situation where 'her partner always asks her about how to make calls and so on, and she is also better at using their TV.' Where one of the older persons in the family was more versed in digital technology use than the others, they appeared to be able to serve, at least to an extent, as warm experts for the others, provided they lived in the same household. They shared their daily lifeworld with those they helped, used the same terminology that the latter did and were to be more readily available than more distant family members when help was needed.

Knowing Me, Knowing You

As already evident by now, intimately knowing the other family members, including their preferred modes of communication and their personal best ways to learn new things, was an essential quality of those acting in the role of warm experts in this study. This was because, as the Slovenian key informant Franc (aged 25) explained it, 'everyone in the family decides for themselves what's the best way for them to communicate with others in view to their knowledge level and their ability to use new technologies.' The older people in the families appeared to be highly aware of the ability of the younger people in them to adapt to their comparatively lower skills and know-how. The Slovenian Mia (aged 25) was one to point this out in her report:

> My grandmother thinks that she can stay away from ICTs, since she belongs to the older generation and is mostly in contact with other older people who do not use many ICTs, either. However, she also thinks that younger people are good at adapting to their seniors, and are therefore able to spend a lot of time talking to them without ICTs.

When the differences in the family members' skill levels and the kind of technologies used within the family were relatively small, one could usually just choose from among the various communication modalities the one that was best suited for the needs and capabilities of others. An example of such a situation was provided by the family of the Italian key informant Melissa (aged 25). As Melissa explained it, in her family even grandparents owned 'more or less recent-model' mobile phones. This enabled Melissa to stay in touch with everyone via mobile phone. However, this she did 'in different ways, depending on the person that I'm contacting'. When someone in the family purchased a piece of new technology or started using a new application, the established manners of intra-family communicating could, accordingly, change, with particular attention again paid to individual needs and capabilities to select a suitable mode of communication. The Italian Alice (aged 23), to take one example, had recently purchased a new smartphone that included many new features, while the others in her family still preferred to text message one another. According to Alice, thanks to her modern phone she could nevertheless adjust her communication behaviour to that of the rest of the family: 'With the new phone, I could adapt myself to the habits of the others: if someone was only writing regular text messages, I did that, too, with that person, and I could do the same with email, WhatsApp, and other online messaging applications as well'.

When the generational gaps in skill levels and the way technology was used were large, the need to adjust one's ways and modes of communication was more pronounced. Keeping the family connected via communication technologies then required that one was ready and able to go back to older modes of communication since those could provide the only way to reach others. Erik (aged 25) from Slovenia referred to this need in his report:

> [T]he common denominator in all our family communication is that we all make an effort to keep communication as easy as possible for those least knowledgeable amongst us. What that means is that, although my brothers, my cousin, and my father are all highly ICT educated, they will all opt for out-dated forms of communication if that makes it easier for my grandmother.

In addition to actual skill differences between generations, the key informants' reports also, explicitly or implicitly, spoke of certain ageist thinking and even practices in the extended families. In families in which young people were less frequently in touch with their grandparents, there appeared to be more stereotypical thinking regarding the latter, for instance. The Finnish key informant Marika (aged 20) indirectly contemplated on this possibility in her report, noting that 'the recipient's age and my own image of his/her technical skills' influenced the means she used for making the contact. She then went on to wonder whether in fact 'older people's interest in new communication possibilities has increased' and whether she should perhaps encourage her elder family members to adopt new technologies. Similarly, the Finnish Paula (aged 26) noted that '[p]eople of different ages employ different apps, so you choose the one that enables the best reach. Age and contactability influence a lot which app you choose'.

Sometimes the personal technical preferences could be so strong that they simply led to a certain way of communicating within the family circle being imposed on also the other members of it. A good example of this was provided by Sabrina (aged 23) from Italy, who spoke of her grandmother's (aged 76) strong attachment to her iPad: if photos were shared in the family, they had to be sent to her by email. The grandmother refused to have anything to do with a computer and lacked a mobile Internet connection. All this was because she 'finds her iPad simpler to use than a PC', and because 'in the iPad mail application she does not have to login every time and then remember her password and username.'

Intimately knowing others makes it also easier to ask and give help. Indeed, the ease of asking was a major topic discussed in the reports by the Slovenian key informants, whose families, as already noted above, were on average less geographically distributed and had more frequent in-person contacts. In them, close relationships and regular face-to-face encounters allowed help to be requested without having to fear losing one's face due to 'digitally disability'. The Slovenian Klara (aged 28) testified to exactly this advantage, writing as follows:

> Older family members don't find it difficult to ask for help and they don't feel that their doing so burdens other family members, as we in the younger generation try to help them to the best of our abilities and be as kind as possible. But it may still be that they secretly feel 'incompetent' about having to do so.

In the key informant reports, asking for help was, however, also thought of as something resulting from sheer necessity: one had in reality no choice other than to rely on younger family members' assistance. The Slovenian key informant Jakob (aged 26) suggested first that the older members of his family perhaps felt that asking for help from others was not that easy, but that '[w]hen they realize that they have no other choice but to ask for it, they do that and are very grateful for the help.' Franc (aged 25), also from Slovenia, had a similar observation: 'My grandparents and parents generally have no problem asking for help, as they are perfectly aware that they can't use ICTs fully without the younger generation's help'. While there was a common agreement that, overall in Slovenia, older people did not hesitate to request help, it was also stressed that the help was given to them without considering it as a big burden. Examples supporting this view included Sandra's (aged 25) grandfather:

My grandfather says that he does not feel capable enough to teach someone else how to use the different functions of a mobile phone, but my parents and my uncle are a bit surer of themselves…. None of the people who help him [the grandfather] accept any payment for it. He knows that he can count on all of his loved ones in case of problems, and he knows that he is not a burden to any of us.

All in all, the role of a warm expert was thus presented as something natural next to all the other family roles: in it, too, those more capable than others helped and provided guidance to other family members in need of such. In Slovenian families, the physical proximity of others made help-giving and help-receiving seem like an organic practice. As Erik (aged 25) put it, '[w]hen it comes to asking for help, none of my family members feel shy about doing so, and they are happy with any sort of help they receive from either one of us.' When the distances between family members were longer and the in-person encounters between them less frequent, as in Finland, help was often provided over the phone or one simply waited until someone came around and provided hands-on help, face to face.

Rewarding but Challenging

The key informant reports, however, also revealed warm experts to not be fully convinced about whether solving technological problems for others was always a good thing from the point of view of the latter's learning prospects. In general, helping other was felt to be rewarding, although acting in the role of the warm expert also entailed challenges and feelings of frustration. For example, Katja (aged 25) from Slovenia made a very clear distinction between what she liked and did not like about helping others: 'Buying new ICT tools is, by far, most entertaining for me, whereas teaching them to use them a bit less so.' She went on to specify what made teaching others less fun: 'Basic questions get repeated over and over again, explaining how to operate any device can take a year or more before they get fully comfortable with it.'

The most frustrating aspect of the warm expert's work appeared to be the fact that it required a lot of time. The older the helped family members were, the more time and effort it took to make them learn things. Moreover, teaching older family member also often required from warm experts that they were willing to be physically present, as suggested, among others, by the Slovenian Marija (aged 25):

Based on a one-week observation period that I've had now, I can say that my older family members learn new ICT operations very slowly and that they need a lot of help with that, and that they need help over longer periods of time, prefer face-to-face assistance, and so on.

Especially, the younger ones among the warm experts in the families considered it tiresome that they had to demonstrate the same things repeatedly, while the learning outcomes might still not be that notable. The following quote by the Slovenian key informant Angela (aged 27) is one of the many (in addition to, e.g. the Slovenian

Franc, aged 25, Mia, aged 25 and Katja, aged 25; the Italian Claudia, aged 21) speaking of this frustration: 'Although I think I've shown her [Angela's mother-in-law, aged 62] how to send an email at least 100 times, she still calls me with problems. Sometimes that annoys me, but of course I still like helping her.'

If having to teach the same things over and over again was sometimes annoying to the warm experts in the families, the slow pace of the learning process could be that to the recipients of the help as well. According to the above Slovenian key informant Marija, the older people in her family 'expressed feelings of dependency, inferiority' when they realized themselves needing help and received help from others. Another Slovenian key informant, Mia (aged 25), noted the following about her grandmother (aged 77) who, she believed, had realized that she was possibly bothering her grandchildren with her requests for help:

> They [the grandchildren] are tired of constantly having to explain the same things again and again. My grandmother has no problems asking for help, but she does sometime get the feeling that she annoys her grandchildren with her repeated questions. She cannot help it, though, as she does not use the functions she needs often enough to remember them.

Reports by some key informants also described certain responsibilities of warm experts that had changed along with the developments in digital technology. Natalija (aged 30) from Slovenia was one to speak of such, describing here how things used to be before:

> When mobile phones first came to the market, it was easy for me to jump into using text messaging, but my parents found that difficult and it took a while for me and my sister [aged 23] to teach them how it was done. Now they find it easier and simply take it for a basic functionality of mobile phones.

After switching from basic mobile phones to smartphones Natalija's parents, however, again began to need more help, although now more with the 'soft contents' of their devices:

> My parents very rarely use any applications on their smartphones. But it sometimes happens that they want to know something about a certain application they have on their phones, and then my sister and I give them a detailed explanation of how it works and why it could be useful for them. But because they don't use them anyway that often, they soon forget what we tell them. Then, after a while, we have the same conversation again, just about a different application maybe. That kind of situations take a lot of energy from all of us, as the process of learning is not easy. Especially if it's necessary to repeat the same thing again pretty soon afterwards.

When it came to teaching technology use to older people, some warm experts had, moreover, found the proverb 'Repetition is the mother of all learning' to not be that reliable in the end. Thus, they had come up with alternative teaching methods. The Finnish key informant Maria (aged 24), for instance, told that members of her family had together written down step-by-step instructions on paper for their grandmother. Before that, the grandmother needed help constantly, with often the same basic things repeated over and over again.

The reports by the key informants Erik (aged 25) in Slovenia, Simon (aged 24) in Finland and Alice (aged 23) in Italy told of another strategy used to facilitate older family members' learning. The three of them relied on encouragement, although sometimes they forced their help recipients to learn by trial and error. Of them, Erik, who himself was the warm expert in his family, described how that could work:

> Sometimes we feel it's better to let someone google the answer to their problem and try to resolve it on their own, rather than giving it to them right away. That might seem a bit harsh to those with the problem, or unnecessary—to my father especially—but I myself feel it's vital for their becoming more independent as ICT users. Doings so has, moreover, brought very good results in our family. These days it might be my father who sometimes comes and teaches me something new, and he shows a lot more interest in technology than before.

When speaking of this strategy, the Finnish Simon described himself as 'sparring partner' in it, rather than a problem solver who would simply do the thing for the other person. In Italy, Alice (aged 23) and her sister had solved the problem with finding the right teaching method by making their parents watch tutorials on the Internet. All these three examples speak of how warm experts, upon the realization that their efforts do not really pay off, often start looking for other ways to obtain learning results and manage their own workload. At times it could also be that such alternative strategies were, in fact, their only options since warm experts do not always have ready answers of their own to the sometimes unexpected or difficult questions directed at them. When that happens, warm experts may then guide the help-seeker in the right direction, as in the case of the Slovenian Aleksej (aged 25) who, hinting of his occasional frustration, explained that:

> I do not like to help with applications I'm not familiar with or those that I don't use myself. The most recent example of those was when my father installed JStock, an application for monitoring shares, and he wanted me to help him use it. In that case, I found a guide online and told him to go look for the answers to his questions there.

Also, the expertise of the warm experts has its limits, sometimes very concretely. When these limits are met, even good intentions and all the willingness in the world to help out are not enough. This had happened to the Slovenian key informant Sebastjan (aged 26), among others: 'There are situations when both my mother [aged 47] and my sister [aged 21] encounter a problem that neither I or my father [aged 50] know the context of, and then they get frustrated when we cannot help them.' The Italian Bruno (aged 27) expressed frustration in this regard, too. He strongly felt that his older relatives tended to overestimate his actual digital skills, only because he worked on a computer daily. Also, the Slovenian Sonja (aged 25) moaned that '[s]ometimes it's hard because both of our parents expect us to know everything and want a reply immediately when they stumble upon a problem'. The Finnish key informant Jenny (aged 25), however, had parents who acknowledged that 'it's often hard for young family members to help [others] with communication devices'.

Compared to early 2000s, warm experts' scope of work has thus considerably expanded, including no longer just help in hardware purchasing and installation, or in convincing others about the advantages of next technology. The work of what we could call the Warm Expert 2.0 increasingly consists of assisting family members with software and programme management and ensuring the functionality of the networked home. When the responsibility for the provision of help in digital technology use is more and more placed on the shoulders of one or two experts in the family, it seems clear that warm experts' personal limits, both technical and mental, will be tested. Caught in a squeeze between high expectations from family members and a constantly evolving personal media and communication technology landscape, the warm experts' relationships with their help recipients do not remain free of intergenerational ambivalence and even conflicts. Young family experts grow frustrated when they are not listened to and their repeated teaching efforts fail to produce long-lasting learning outcomes (as, e.g. in the family of the above Klara, in Slovenia). Correspondingly, older family members may become frustrated when realizing that all the help in technology use that they need is not readily available through warm experts (as, e.g. in the family of Sebastjan, also in Slovenia).

References

Bakardjieva, M. (2005). *Internet society: The internet in everyday life*. London: Sage.

Barnard, Y., Bradley, M. D., Hodgson, F., & Lloyd, A. D. (2013). Learning to use new technologies by older adults: Perceived difficulties, experimentation behaviour and usability. *Computers in Human Behavior, 29*(4), 1715–1724.

Dolničar, V., Grošelj, D., Hrast, M. F., Vehovar, V., & Petrovčič, A. (2018). The role of social support networks in proxy Internet use from the intergenerational solidarity perspective. *Telematics and Informatics, 35*(2), 305–317.

Selwyn, N., Johnson, N., Nemorin, S., & Knight, E. (2016). *Going online on behalf of others: An investigation of 'proxy' internet consumers*. Sydney: Australian Communications Consumer Action Network.

Chapter 6
Digital Housekeeping

Abstract At this point of the book, the concept of 'digital housekeeping' is introduced and applied in the context of the overall investigation. Based on existing research, digital housekeeping tasks and responsibilities are broken into three subcategories to facilitate analysis: hardware installation and configuration, digital content and software management, and transfer of knowledge within the family. In the Finnish, Italian and Slovenian families in this study, digital housekeeping tasks, especially those related to software, were typically assigned to the young warm expert(s) in the family. In hardware-related matters, the family's digital housekeeper could also be someone else, such as the father of the family. The chapter concludes with the suggestion that a family's digital housekeeping tasks and responsibilities are likely to become reorganized and redistributed as its members grow older, it changes shape or its older family members develop more digital skills.

Keywords Digital home · Digital housekeeping · Home maintenance · Household chores · Housekeeping · Technology purchases · Warm experts

The previous chapter discussed how one becomes, and what it means to be, a warm expert in digital families in which family members make extensive and varied use of digital technology even if their individual skills levels as well as their modes and styles of using that technology may be very non-uniform. In this chapter, the attention is turned to the tasks and responsibilities involved in the maintaining of the digital home. How big a role do warm experts play in ensuring that digital devices and applications work properly in the digital home? While the issue was already touched upon in the previous chapter, a more detailed analysis is presented here, focusing on the digital housekeeping activities to have emerged as a consequence of the digitalization of the domestic sphere of life.

The concept of *digital housekeeping* refers to all the tasks, chores and responsibilities involved in the maintenance of the networked home's functioning. In previous research, digital housekeeping tasks have been broken down into three main subcategories: hardware purchases and configurations, software and application management, and transfer of knowledge (Kennedy, Nansen, Arnold, Wilken,

© Springer Nature Switzerland AG 2019   75
S. Taipale, *Intergenerational Connections in Digital Families*,
https://doi.org/10.1007/978-3-030-11947-8_6

& Gibbs, 2015; Tolmie, Crabtree, Rodden, Greenhalgh, & Benford, 2007). The discussion in this chapter is organized so as to reflect this categorization, in order to better be able to document the various aspects of digital housekeeping observed in the three countries in this study.

Housework Meets Digital Technology

The concept of digital housekeeping opens a fresh and modern vantage point for the study of the division of housework within the family. For it, besides a rich body of qualitative and historical research on the division of housework, much of which specifically focuses on sex segregation (e.g. Jackson, 1992; Oakley, 1974), also the established field of time-use research looking into the division of housework at the household level can be drawn upon (e.g. Gersbuny & Sullivan, 1998; Hook, 2010; Oinas, 2010). Time-use diary data sets, coded in uniform time-use categories (e.g. Harmonised European Time Use Survey [HETUS]), have enabled international comparisons and time trend analyses regarding the proportion of household chores performed by men and women, respectively.

In the established time-use categories and classifications, digital media and communication technologies fall under free-time activities and are considered as belonging to the domain of mass media consumption. However, even casual observation suggests that digital technologies also have brought with them new kinds of maintenance and meta-work not limited to entertainment and pastime functions only. Sustaining the functionality of the home and the daily life more and more entails spending time on tasks such as installing, configuring, pairing and updating various devices, programmes and applications. What all that could mean for one's daily life is in the following quote given an example of by the Italian key informant Enrico (aged 24):

> My grandfather instead turns to me for the deletion of the call log, the checking of the messages to be read, and some of the routine maintenance work on the PC. I'm also the one telling him when there are emails that are important to him. I do the same with my parents as well. My dad sometimes tells me to go check his messages in case there's anything he should read, while my mom sometimes asks me to change some setting on her business smartphone or to help her do some online banking thing or the like…. As regards our home computer, it's my job to keep the antivirus programme up to date and ensure that all the different programmes work properly. Also my aunt turned to me for advice when she was buying a new smartphone, and also afterwards, to get detailed instructions from me on its use, especially how to configure Internet access and transfer photos from it to her notebook.

Just as with any other housework, also digital housekeeping appeared to evoke ambivalent feelings about the distribution of responsibilities among family members. Some key informants claimed the responsibilities to be fully and evenly shared in their families, arguing, like the Slovenian key informant Tina (aged 25), that 'the responsibility for the proper functioning of our devices and programmes is evenly distributed amongst us all in the family; no one is particularly in charge'. Claudia (aged 21) from Italy explained, along somewhat lines, that the way digital house-keeping tasks were divided in her family had come about 'totally spontaneously and in a most natural way' and that '[t]here have never been any arguments or friction between us about the way these roles are distributed'. Other key informants like the Slovenian Angela (aged 27), however, went on to complain that, in comparable situa-tions in their families, there were times when it felt like no one had or was prepared to assume the responsibility for making sure that the digital devices functioned, fixing them when there was a malfunction.

Indeed, it was quite often the case in the families that when problems or unexpected situations arose, responsible persons on hand were few and far between. As also this research indicates, not much progress had been made in ensuring a fair division of technology-related housekeeping tasks over the years. Quite the contrary, compared to how things had been before, the situation had even become less clear and less established. As the grandmother (aged 79) of the Slovenian Boris (aged 26), for instance, bemoaned, in the past 'everyone knew who was in charge, but today it's no longer so'. Media and communication devices had become highly personal, leaving the household equipment that everyone used in no-man's land, as it were. In the next sections, the current state of digital housekeeping practices in the three countries is taken up in more detail, looking at how the tasks in question were divided in the extended digital families in them.

Hardware Purchases and Maintenance

It should come as no surprise at this point that the main responsibility for digital housekeeping in the digital families also in this study fell into the hands of the warm experts in their midst. Typically, these were part of the younger stratum of the family and were considered as having strongly influenced the purchase of the digital hardware (as in the families of, e.g. the Slovenian Angela, aged 27, and Klara, aged 28; the Italian Mario, aged 24, and Enrico, aged 20) and being thus also responsible for their maintenance (as, e.g. in the families of the Slovenian Anita, aged 28, and Tina, aged 25; the Italian Marco, aged 24).

The young warm experts' digital housekeeping role was highlighted in the key informant reports particularly well when the question was about family practices aimed to secure the proper functioning of technological devices in the family. The Slovenian Petra (aged 25), for example, told as follows: 'When there is a problem, people in our household turn to the person in the family they know can fix it: they turn to either my brother or me and want us to make their gear work the way it should'. Also, in the Italian key informant Emilio's (aged 30) own family and parental family, it was clear who had the responsibility for digital housekeeping, even if the situation differed between the two contexts in this regard:

In my own house [where I live with my girlfriend] I am the one handling all the technical aspects and making sure the information technology works, although my girlfriend and I, we are both very knowledgeable about how to use ICTs.... However, where my parents and my siblings live it's my father whom the rest of us turn to when it comes to technical stuff: he is seen as a kind of consultant whenever there's a malfunctioning machine to be fixed or a new purchase to be made.

The idea, expressed also by Emilio in the context of his parental family, that parents are the family's decision-makers was deeply rooted in the minds of the key informants and their siblings. When it came to major decisions such as those concerning the purchase of shared household technologies (a new digital television set, broadband Internet, etc.), it was frequently underlined that even if children were almost as a rule always consulted and listened to, the final decision was not theirs but their parents', and even then more typically the father's rather than the mother's. The Finnish key informant Sara (aged 25) supplied an example of this in her interview report. Her brother very firmly took their parents' opinion to be decisive when buying appliances for the home. As Sara clarified, this brother did, though, always extensively discuss any technology purchases with his father first before the transaction was made, while the mother of the family was also heard, to obtain her opinion on whether the new equipment was in the end really needed in the family or not. Sara's own interpretation of all this was that 'my parents' opinion is not necessary final and absolutely decisive for my brother, but it affects him, even if that might be mostly subconsciously. By saying so, she suggests that deferring to one's parents, as in her brother's case, was more of a cultural norm than any actual determining factor. A somewhat related view was put forth by another Finnish key informant, Laura (aged 29), who stated that 'the parents talk with their youngsters before any equipment purchases are made because the young people know more, but when the parents pay for the purchase, they are also the ones to make the final decision'.

Along these lines, many informants made a clear distinction between formal purchase decision and provision of information influencing or leading to that decision. It was argued, for instance, that while it was the parents who stood for the former, the latter role was young warm experts' purview. In the following quote, the Finnish key informant Carla (aged 23) provides a case in point:

My father thinks that it's he who makes the decision to buy something, but then it's my brother who decides what kind of device we are actually going to buy. So the decision whether, for instance, we should or shouldn't get a computer is made by my father, but what kind of computer it's going to be, in terms of its technical properties, is then decided by my elder brother, who's in our family the one who knows most about ICTs.

Thus, even when the older family members indeed paid for the purchase, the young warm experts had considerable influence in practice on what kind or type of hardware would end up being acquired. As the Slovenian key informant Alexander (aged 24), for instance, explained, '[w]hen it comes to technology purchases, my father [aged 58] and my mother [aged 44] always ask me or my brother [aged 18] for advice first, since it's the two of us who are the most knowledgeable about these things in our family'. Slovenia, however, was not unique in this respect, as, for example, Alessandro (aged 20) and Claudia (aged 21) in Italy could testify:

When there is new ICT equipment that we need to buy in my family, it's definitely my sister and I who are our experts for it. That means that our roles drastically change, if not get totally reversed, compared to the usual situation. The decisions are, in other words, de facto taken by the children who describe and explain to their parents the differences between the various products. (Alessandro)

The purchases of new devices that we do are most often influenced by the younger ones amongst us, as they are far more advanced and experienced in information technology than their parents. But it's always possible to find a compromise between the two parties. (Claudia)

Even though parents' de facto decision-making power in this study, due to their limited knowledge of new technology, was at least in part apparent only, both parents and their children were nevertheless aware of the former's ability to actually control family spending. As the father of the Finnish key informant Carla (aged 23) put it, 'also the wallet decides'. The Slovenian key informant Franc (aged 25), too, paid attention to the cost factor in his family's purchases in this regard, explaining that, in their case, the more expensive purchases were decided upon jointly by the children and the parents. As he noted, '[s]ince the purchases of new ICTs typically cost a bit more and are therefore most often paid for by the older members of the family, we nevertheless make the decisions about them jointly, based on past experience and needs'. Similar observations were put forth in Finland and Italy, too. The Finnish key informant Marika (aged 20) summed up the situation in her family by stating that, in hardware acquisitions, also the household's current financial situation mattered. In Italy, the uncle of the key informant Emma (aged 22) stressed how 'the economic dimensions of those [purchase] decisions are actually controlled by the adults, since they decide how much will be spent'.

All in all, the act of purchasing new technological devices in the digital families studied emerged as a deliberative and intergenerational. For the decisions to go smoothly and be made un-conflictually, confidential relationships within the family appeared to be necessary. The Slovenian key informant Boris (aged 26) was one to highlight the importance of trust in his family's decision-making in this regard. As he explained it, this trust became especially significant when 'the person getting to make the purchase decision in the family is the one who is most skilled at using the technology, while the others must simply believe that it's a good decision'. In fact, according to Boris, this was so, in particular, when 'the purchase concerns technology that affects all members of the family, like the Internet, television, or telephone service provider, for instance'. Also, the Italian Melissa (aged 25) underlined how in her family technology purchases resulted from joint decisions. As she explained, when her grandparents, for instance, grew interested in some particular device, they first solicited advice concerning it from others, after which it was then decided 'together which option is the best for the family to choose'.

What is important to note here, however, is that not all hardware appeared to be similar in this respect. There was, for example, a clear difference between purchasing low-priced, personal communication technologies and more expensive household technologies. While the importance of intergenerational dialogue was in both cases emphasized, particularly when the question was of a purchase involving more expensive technology for the family's use, several of the informants also stressed that everyone should at the same time be responsible for their own personal technology purchases and maintenance. This idea of personal responsibility and the need for self-sufficiency was particularly notable in families whose members were financially independent of one another. The Slovenian key informant Jakob (aged 26) provided one example, describing as follows what he had found among the adults with a steady job or another source of regular income that he had interviewed for this study:

> Everyone of them decides for themselves about their purchase of their personal ICTs. If it's a bigger purchase, then they consult with their partner and also their children and other family members whom they perceive as the most knowledgeable about that stuff, being thus able to offer the best advice, and after that they all arrive at a joint decision.

Having personal technologies thus also meant accepting personal responsibility for the devices' functioning. Among those underlining this fact was the Slovenian key informant Petra (aged 25), who stated that '[w]hen it comes to mobile phones, each one of us is responsible for her or his own gear'. Similarly, the Finnish Carla (aged 23) pointed out that 'I feel like if people have their gadget in a personal, and not common, use, then they, at least to some extent, are the ones with the responsibility for it, on their own. So that if the thing stops functioning, you usually try to solve the issue by yourself, before asking others for help'.

Indeed, in this study, this idea of personal responsibility associated with independent purchases was most often put forward by the Finnish respondents. In Finland, we may recall, financial independence is typically achieved at a relatively young age. As the 21-year-old Julia, for example, noted, '[i]In this golden era of hire purchase, even students like us can easily afford laptops and smartphones'. Yet, as another Finnish respondent, Laura (aged 29), reminded, the level of independence in such decisions varied based on the person's disposable income from one life stage to the next: 'Everybody takes their own hardware purchase decisions based on their current personal needs and their current financial situation; everybody these days has personal devices, no longer just, let's say, a shared family computer'. Also, Mary (aged 26) in Finland spoke of how one's current life stage and living arrangements influenced the degree to which one's hardware purchases could be decided upon independently:

> When planning to buy equipment, grown-up children make their own decisions, while a family living together in the same home—father, mother, two youngish children—makes hardware and software purchases together, with older children making interventions.

Interestingly, some key respondents nevertheless stressed their parents' relative independence as decision-makers, contrasting them to their grandparents. For example, the Finnish Simon (aged 24) summarized the findings from the interviews he had conducted as follows:

> When it came to hardware purchases, the independent role of the parental generation was emphasized more, whereas for grandparents, younger family members were more often enlisted as advisors and participants for the decision making when acquiring, for example, a computer.

Grandparents' lower degree of independence in their technology purchases was also described by the Italian Sabrina (aged 23), who reported that her 'grandmother leaves the choice about the ICTs to be purchased to my parents or, alternatively, my brother'. Here, however, one should note that technological independence does not, in general, and also in this study did not, decrease in a linear fashion with age. Some of the key informants for this study had relatively young parents (in their mid-50s and early 60s) who had outsourced all the decision-making in technological matters to their descendants. One such person was the Slovenian Angela (aged 27):

[M]y mother [aged 53] and my mother-in-law [aged 62] do not participate in the decision making about the new technological equipment to be acquired. They do not even have a say in the purchase of their own mobile phones, which are not smartphones.... And if my father [aged 63] needs a new one, he lets my brother know what it is that he needs and how much money he is willing to part with for it, after which it's then my brother who goes out and finds something fitting the need.

All in all, the division of digital housekeeping tasks in new hardware purchases and maintenance thus varied across families and countries, while one thing remained constant: the possible purchases were typically highly dependent on older family members' level of digital engagement. The more 'digital-ready' the key informants' parents were, the more independently they made their technology purchase decisions (and the actual purchases). Major (read: expensive) household technology purchases, such as of TV sets and laptop computers, were more often than smaller ones, such as a smartphone, discussed and completed together with other family members. However, the one common feature in all of these situations—and this was so in all the three countries studied—was that young warm experts were in every case given the main responsibility for the proper functioning of the hardware ultimately brought home.

Software and Application Installations

A task that even more often and more conspicuously than purchasing and maintaining hardware was left for the young warm experts in this study to fulfil was that of managing software and applications. According to key informants, all that which was 'hiding inside' their devices or could be installed on them represented a terra incognita for many of their late-middle-aged parents. As the reports they submitted revealed, especially key informants' parents frequently sought advice from their young warm experts when encountering a problem using their smartphone. This was the case, for instance, with the parents of the Italian key informant Alice (aged 23):

My parents only use these apps because I and my sister help them do that. At the beginning, they were only able to use their phones in the offline mode. When my parents discovered new technologies and began to understand the benefits of using them, they did for some time explore them on their own, but right now they're constantly coming to me and my sister again for help, as sometimes they just keep forgetting even the basics of how their apps work.

Also, the Slovenian key informant Anton (aged 29) reported himself to be the person whom his 'parents turn to when they need help using their mobile phones or PCs'. Both of his parents had bought their mobile phones without any help from their children, but afterwards they needed advice in the use of applications and with certain smartphone functions. Anton explained that he had 'set up email accounts for them [parents'] along with some functions like speed dialling and some other mobile phone settings'. In other words, Anton's parents were confident enough to make their own technology purchase, yet felt unsure when it came to what was hidden from sight inside the sleek exteriors of their phones. Somewhat similarly, also the Slovenian key informant Katja (aged 25) described that, because her brother, who was the other warm expert in the family, was often away from home, she had ended up being the sole person at home to help her mother to, for example, instal new applications like Viber and keep instructing her grandmother about how to send text messages.

Quite often, the key informants' parents also required help in installing computer programmes. Because of their lower engagement with smartphones, many of them had kept using stationary communication devices, such as desktop computers. According to the Slovenian Franc (aged 25), in his family 'parents are for the most part only taught how use computers—how use of the different applications, programmes, etc., in them'. He then specified having 'mostly taught them how to use Facebook, and some slightly more advanced Microsoft Word functions, and how to instal new software if needed'. In the family of the Slovenian key informant Veronika (aged 27), it was her younger brother 'who solves the issues with software, like anti-virus programmes' for their parents and the rest of the family.

The fact that the parents needed and received help with software installation and updates did not, however, mean that they would not have liked to learn new things and new skills. Quite the contrary, many of them were eager to learn new programmes and applications. According to the interviewees, this, to be sure, then also added to the young warm experts' already fairly long to-do lists. The Finnish key informant Rita (aged 34) wrote about her father who had grown interested in a new operating system, the Linux-based Ubuntu, after having seen her Rita and her partner happily use it. At the time of the interviews, Rita's father was contemplating on becoming a Ubuntu user himself, with Rita anticipating him to certainly need help in using it, and that 'if that happens, he will be needing a lot more help from my husband with all the software upgrades, as with Ubuntu at least some coding skills are needed for that'.

Older people's dedication to desktop computers often complicated the work of the young warm experts, who found mobile devices much easier to use as well as more reliable. In Italy, Elisa (aged 26), for instance, described her brother (aged 30) and her uncle (age N/A), who were the warm experts in the family, as 'the reluctant maintainers of our smartphones, PCs, and other communication devices' who often expressed 'frustration about how desktop applications are more prone to fail in less expert hands like those of my parents'. Her brother kept attempting to rid himself of those unpleasant digital housekeeping tasks, towards which purpose he pushed the parents to use apps installed on their phones rather than on their PC, since smartphone apps are simpler to restore and far more tolerant of misuse.

The same way as when making purchase plans for a new piece of hardware, intergenerational negotiations were also entered into the digital families when family members needed or wanted to make a choice between different applications. In her report, the Finnish key informant Ella (aged 24) described one such situation, showing how differently the matter could be handled in her family, depending on whether the question was of purchasing new hardware for her younger brother or deciding on which applications he should use:

> My stepfather and my mother make often purchase decisions for my little brother [aged 11], but when it's about which applications to get, we sometimes first discuss the matter amongst us siblings, and then mull it over with my mother and my stepfather, trying to together figure out which ones my kid brother might be able to ready to start using.

The quote touches upon some interesting aspects of intergenerational relationships in digital families. First, parents as the financiers of hardware purchases might have the final say regarding them, but in the case of free-to-download mobile equipment applications they did not, or were not able to, leverage the same power. Second, since they tend to be less familiar with the diverse mobile applications than the younger members of their families, they can but trust in the information provided them by the young warm experts. Third, and perhaps most interestingly, the quote also reveals a normative aspect of intergenerational solidarity relating to the use of digital technologies (for more on that, see Chap. 8). Although the key informant and her siblings here might have had just outright dismissed their parents' opinions, letting their younger brother instal new applications for himself without consulting with his parents first, they nevertheless decided to ask for their parents' opinion, too. Indeed, they felt obliged to do so.

In a few key informant reports, however, a different aspect of normative expectations was brought up. This related to what we could call 'mothering'. As the reports showed, digitally skilled mothers sometimes considered themselves responsible for ensuring the proper functioning of software and applications in the family. Here, the traditional role of mothers as the maintainers of the home and domestic social relationships, including the dimension of family communication and care provision, was extended to software care, too. In one case, from Finland, the key informant Teresa (aged 24) described the situation as follows:

> My mother [aged 58] herself feels that she's the technical expert in the family, and that she should take care of all the downloading and purchases. But also the grown-up children can be asked for advice on these things, and sometimes, if it's about installations, the younger ones, too.

Also, others, like the Italian key informant Emma (aged 22), reported their mothers as feeling themselves responsible for ensuring the smooth functioning of the digital technologies at home. In Emma's case, while her grandmother 'blindly relies on her children for any problem, since she would not be able to either identify or fix any of them', it was particularly her mother who 'considers herself to be the one with the responsibility, fallen on her due to her high personal skills, for the correct functioning of the Internet, the online connection, and the phone plans'. As Emma's report suggests, with the rise of women's digital skills, digital housekeeping tasks may quietly end up becoming included in the already wide array of domestic chores that women are expected to handle and manage anyway.

Knowledge Transfer Within the Family

The third dimension of digital housekeeping relates to the transfer of knowledge about new digital media and communication technology (see, e.g. Kennedy et al., 2015; Tolmie et al., 2007). The existing research paints a picture of two-way knowledge transfer between family generations, one in which young people convey new knowledge and know-how to their parents and grandparents while the parents transfer certain knowledge, yet about very different matters, to their children (see Chap. 3). The Italian key informant Irene (aged 24) highlighted this difference particularly well when comparing the views of her young and middle-aged interviewees, naming these as the Young Group and the Middle Group, respectively:

> We can speak of a mutual exchange of knowledge: those in the Young Group are, for example, more practical and experienced on the application side of ICTs, while those in the Middle Group know better the technical side of these tools. In this way the people in the two groups complemented each other's different abilities. If something doesn't work properly in one of the classical technologies like the PC, it's the Middle Group that deals with it first, whereas if it's about some more recent technologies, it's every man for himself.

The mutual exchange of knowledge that Irene talks about here followed largely the same pattern that held for the hardware and software-related digital housekeeping tasks in families. As children, on average, appear to be more knowledgeable about mobile applications and computer programmes than their parents and grandparents, they pass their knowledge related to these areas to the older people in their families. As the Finnish key informant Sara (aged 25) explained:

> Younger interviewees tell that their parents use several different applications. These interviewees taught them in their use, advising them and showing them also concretely how a device or an application works. My brother told me that this instruction sometimes also takes place so that the parents just watch the younger people use the kind of apps they are interested in and want to start using.

In general, the knowledge younger people passed on to their parents and grandparents was offered in the form of concrete pieces of advice, such as about how to use a certain device, application, programme or service (as, e.g. in the families of Carla, aged 23, and Maria, aged 24, in Finland; Silvia, aged 25, and Marco, aged 24, in Italy; Sonja, aged 25, in Slovenia). The Italian Silvia (aged 25) had taught her great-uncle (aged 82) 'how to save a document, delete mail, and print stuff', while Anton (aged 29), from the same country, reported himself having 'taught everyone in my family how to use Google Hangouts, because they all have smartphones with the Android operating system that has the Hangouts pre-installed in it'.

In the same vein, the Finnish Rita (aged 34) maintained that she, her husband and her sister had taught their parents everything these knew today about how to use personal communication technologies and computer programmes. As she further stated, '[m]y father asks my husband for help with Windows and some other programme updates at least three times a year'. Then, turning the attention to her own daughter (aged 5), she went on to also describe the kind of knowledge that her own parents tried to pass on their next generation:

The same way, with my husband, we are slowly beginning to teach our daughter how to use telephone and send and receive text messages. We might get her a phone of her own when she starts school. My own parents, on the other hand, are trying to teach my sister, as diplomatically as possible, of course, about what information and images on Facebook might be worthwhile to share and with whom.

Rita was far from the only one to report about how, in digital families, parents' efforts to pass on to younger generations what they knew about digital technology and its use mainly involved notions about the 'right' and proper ways of using that technology and its pertinent applications. Other informants in other countries confirmed the validity of this observation in also their cases. The Slovenian key informant Klara (aged 28) did so when discussing what should and what should not be shared online:

The older family members teach the younger ones about how it's not wise to share your personal photos, events, and data with the broader public on social networks. My mother [aged 56] is very adamant about this, always telling me and my sister that we shouldn't share our personal photos with others on Facebook and other social networks.

Willingness to engage in this kind of knowledge transfer might, however, also been seen as a mere extension of parents' general desire to protect their children and raise them in a safe environment. Yet, at the same time, the above quotes also speak of parents' relatively disadvantaged position in today's world of digital technologies. The key informants' late-middle-aged parents had never in their own childhood and youth encountered the kind of risks today's digital media and communication entail, and hence lacked a point of reference for how to deal, as parents, with those risks and the exposure to them. Because of that, they were largely dependent on secondary information and knowledge about appropriate parental strategies, often acquired from the media.

To summarize, this chapter on the intergenerational dynamics of digital house-keeping practices has shown how in Finnish, Italian and Slovenian families digital housekeeping tasks and responsibilities tend to be typically assigned to the young warm expert(s) in the family. While this was especially clearly so when those tasks involved attending to software-related problems or questions, in hardware-related matters the family's digital housekeeper could also be someone else, such as the father of the family. Overall, we can expect digital housekeeping tasks and responsi-bilities to become reorganized and redistributed when family members grow older, the family changes shape or older family members gain more digital skills. Comple-menting recent research that has found digital housekeeping to also involve certain duties facilitating intergenerational cooperation (Fortunati, 2018), this chapter sug-gests that this cooperation among other things empowers younger family members, consolidates family connections and enhances solidarity across generations.

References

Fortunati, L. (2018). How young people experience elderly people's use of digital technologies in everyday life. In S. Taipale, T.-A. Wilska, & C. Gilleard (Eds.), *Digital technologies and generational Identity: ICT usage across the life course* (pp. 102–118). London & New York, NY: Routledge.

Gersbuny, J., & Sullivan, O. (1998). The sociological uses of time-use diary analysis. *European Sociological Review, 14*(1), 69–85.

Hook, J. L. (2010). Gender inequality in the welfare state: Sex segregation in housework, 1965–2003. *American Journal of Sociology, 115*(5), 1480–1523.

Jackson, S. (1992). Towards a historical sociology of housework: A materialist feminist analysis. *Women's Studies International Forum, 15*(2), 153–172.

Kennedy, J., Nansen, B., Arnold, M., Wilken, R., & Gibbs, M. (2015). Digital housekeepers and domestic expertise in the networked home. *Convergence, 21*(4), 408–422.

Oakley, A. (1974). *The sociology of housework*. New York: Pantheon Books/Random House.

Oinas, T. (2010). *Sukupuolten välinen kotityönjako kahden ansaitsijan perheissä*. Jyväskylä studies in education, psychology and social research (402). Jyväskylä: University of Jyväskylä.

Tolmie, P., Crabtree, A., Rodden, T., Greenhalgh, C., & Benford, S. (2007). Making the home network at home: Digital housekeeping. In *Proceedings of the Tenth European Conference on Computer Supported Cooperative Work*, 24–28 September 2007, Limerick, Ireland (pp. 331–350). Dordrecht: Springer.

Chapter 7
The Big Meaning of Small Messages

Abstract Here, instant messaging as a mode of everyday communication in digital families is taken up for examination. We look, in particular, into the qualities that make WhatsApp an attractive communication tool for extended families: it allows both one-to-one and one-to-many interactions and provides multiple modalities for intergenerational family communication (voice, text, photos and videos). Empirical evidence and qualitative data collected in Finland and Italy in 2014–2015 are drawn upon and analysed in advancing the argument that the success story of WhatsApp in the family context is related to way it enables reaching the whole family at once and promotes 'phatic communion' via small messages.

Keywords Extended family · Instant messaging · Intergenerational relationships · One-to-many communication · Phatic communion · WhatsApp

This chapter takes up instant messaging for examination as a mode of everyday communication in digital families. The investigation focuses on one particular communication application, WhatsApp, which, at the time of the data collection for this study, was one of the most popular instant messaging applications in many countries (O'Hara, Massimi, Harpe, Rubens, & Morris, 2014). This was the case also in Finland and Italy. What makes WhatsApp an attractive communication tool for extended families is that it allows both one-to-one and one-to-many interactions, and provides multiple modalities for intergenerational family communication (voice, text, photos and videos). The question is raised whether, to what extent, and in which ways small messages exchanged via WhatsApp might contribute to the sense of social coherence in extended digital families. The discussion draws upon empirical evidence and illustrations derived from qualitative data collected in Finland and Italy in 2014–2015. The chapter also briefly considers the reasons why, at the time of this study, Slovenian families, compared to those in Finland and Italy, had far less enthusiastically embraced the possibilities offered by the WhatsApp application.

The analysis presented in this chapter was originally published in Taipale and Farinosi (2018).

WhatsApp's Growth and Success

WhatsApp is an instant messaging application that runs on mobile communication devices equipped with an Internet connection. WhatsApp allows sending text, picture, voice and video content to either one person at a time or several persons participating in chat groups. In 2017, after the data collection for this book had already been completed, a new feature was introduced in WhatsApp that allowed users to post customized photos and videos timed to automatically disappear after 24 h. The application can be categorized as either a real-time or a near-real-time communication tool.

Other prominent (and popular) features of WhatsApp are that it enables the user to follow the delivery of the message and see when one's contacts are available and when they themselves are busy typing messages. To indicate that a sent message was successfully delivered, a check mark will appear next to it, while two check marks, of varying colour, tell that it has been received and read. Similarly, WhatsApp shows whether other users in one's contact list are currently online and, if they are not, when they have last logged in; this last seen timestamp feature, however, can be disabled by the user. Research has already shown this micro-scale peer monitoring to be commonly used to check the availability of others without, however, any actual intention of contacting them (e.g. Karapanos, Teixeira, & Gouveia, 2016; O'Hara et al., 2014).

Released in 2009, WhatsApp's worldwide popularity has increased rapidly ever since. According to Statista (2017), the total number of WhatsApp users multiplied more than sixfold over the last four years or so, going up from 200 million in April 2013 to 1.3 billion by July 2017. The quality of the available user statistics is, however, somewhat variable, as data is, for instance, not available for all countries and, for those that it is, not always comparable. In any case, what seems clear from overtime comparisons is that the number of the application's users has constantly increased and keeps increasing still today.

Of all the Nordic countries, as the AudienceProject (2016) report shows, WhatsApp was clearly the most popular in Finland in 2016. There, it ranked as the number one social media tool overall, while failing to make it anywhere near the top in any of its Scandinavian neighbours. In the last quarter of 2016, 68% of all Finnish smartphone owners reported themselves using WhatsApp. Finns were, however, also very busy users of the application, with 49% of those with using it claiming to do so several times a day and 29% every day. In comparison, the corresponding figures for Sweden in the same time period were 25 and 16%, respectively. Moreover, as the same report shows, Finnish women were slightly more frequent WhatsApp users than Finnish men (42 and 32%, respectively), and WhatsApp was the most popular social media application in all age groups in the country. Its penetration rate in the country varied, being the highest among those aged 15–25 (70%) and the lowest among those aged 56 or over (18%). The increase in the numbers has indeed been remarkable in terms of speed and the sheer size, as just two years earlier, in 2014, no more than basically one in every third Finn (37%) reported using WhatsApp (Taloustutkimus, 2014). In the other two countries in this study, the penetration numbers were not as readily available (for Slovenia, no reliable statistics were available in general). According to a Deutsche Bank estimate in 2015, however, the penetration rate of WhatsApp among Italian smartphone users was 68% (Stern, 2015).

Family Instant Messaging

A glance back at history reveals that, for a good while, online instant messengers remained a communication media utilized mainly by teenagers for peer-to-peer communication and young adults for work-related interaction (Bouhnik & Deshen, 2014; Grinter & Palen, 2002; Johnston et al., 2015; Lenhart, Rainie, & Lewis, 2001; Nardi, Whittaker, & Bradner, 2000). Also, recent studies point to children's preference for communicating with their peers, not parents, through mobile and social media tools (e.g. Nag, Ling, & Jakobsen, 2016). Even if most of young people's mobile communication might then be with their peers, it does not mean that they would systematically exclude their parents—or even grandparents—from their instant messaging activities, however. As a matter of fact, the ways in which WhatsApp communication is creeping into the everyday life of extended families is still an unexplored territory.

The majority of the relevant research conducted in the area thus far deals with the gratifications of instant messenger and other social media tool use (e.g. Ling & Lai, 2016). Church and de Oliveira (2014) compared the way people use SMS and WhatsApp in Spain, employing both qualitative and quantitative methods. What they discovered was that, among the Spaniards aged 20–60 that they studied, WhatsApp was strongly associated with immediacy, a sense of community and free use, and that these were considered as its main gratifications. On the other hand, text messaging was still felt to be more reliable and entail fewer privacy concerns. In the United Kingdom, O'Hara et al. (2014) studied WhatsApp use among Britons aged 17–49 who came from a variety of occupational backgrounds and included both persons living alone and couples. For the group they studied, WhatsApp was frequently seen as a means enabling one to 'dwell' with others: it was constitutive of the kind of commitment and faithfulness characterizing social relationships, in general, and served the needs of social bonding more than any functional purpose of merely exchanging information.

The migration of instant messaging from desktop computers to smartphones has diversified the socio-demographic profile of service users. Smartphones, with their pre-installed applications and easy-to-use application stores, have introduced instant messengers to an ever-wider group of potential users. They have not only added mobility to instant messaging communication but also extended the overall range of available modalities from text-based messages (as in IRC and AOL's Instant Messenger) and voice calls to photos, voice messages and Internet calls (see, e.g. Baron 2010; Ling & Baron, 2007). This very ability to choose from among many different modalities is what makes WhatsApp and other instant messengers like it suitable tools for connecting people with different communicative preferences. In offering something for everyone, WhatsApp allows users to adapt to one another's communication preferences, habits and manners and, by so doing, helps family members overcome social differences between family generations in that regard.

In extended families, instant messengers need to be positioned into the matrix of the intricate parent–child relationships, which also reflect children's mutually contradictory needs for autonomy and parental care. As previous studies have shown, mobile communication devices, in general, serve both ends here: they serve to maintain an 'umbilical cord' between children and parents and function as a medium enabling children's greater degree of independence (Ling, 2007). Somewhat along the same lines, Ribak (2009) has looked at the mobile phone as a kind of transitional object in family life, one that can be viewed as a materialization of the parent–child nucleus around which the relationship between the two is continuously communicated, negotiated and redefined. In the family context, however, the social roles of parent and child are also easily inverted. In parents' use of mobile communication tools, also their dependence on their children's technological assistance and caretaking is manifested (Taipale, Petrovčič, & Dolničar, 2018). The same concerns the relationship between grown-up children and their ageing parents, whose dependence on others, in general, only increases with age.

The ability to sustain and nurture family connections from afar that mobile communication has meant has prompted researchers to argue that new social media and digital communication technologies have given rise to 'networked families' or new relational families (Horst, 2006; Lim, 2016; Madianou & Miller, 2011; Rainie & Wellman, 2012; Wilding, 2006). Yet, there are only a handful of studies exploring the actual ways in which families use mobile instant messengers and their group chat functions, in particular, to stay connected. One of them is by Rosales and Fernández-Ardèvol (2016), who have showed how in Spain, where WhatsApp is commonly used across all age groups, the way smartphones are used, rather than being only based on age-differentiated skills, typically reflects the users' interests in technology use and communication needs that change as one grows older. Elsewhere, Siibak and Tamme (2013) have studied how various web-based communication channels are used in Estonian families, finding these new communication tools to be appreciated by families, especially for their ability to offer a sense of closeness among family members. This function of theirs was especially valued among older family members who lived apart from their children, while for younger people also their ability to act the same way in one's peer relationships was important.

Siibak and Tamme (2013) went on to argue that web-based communication technologies serve family relationships when family members live in the same household. Although the mobile devices and applications the families in their study used were highly portable, they were to a notable extent also deployed to coordinate activities and share information in relatively close proximity of other family members, individuals who were sometimes even located within the same household and including in entirely non-mobile situation (cf. Fortunati & Taipale, 2017). Indeed, based on their findings, too, it appears that the newer forms of social media can support group and small community interaction to a higher degree than older one-to-one technologies and earlier social networking sites, in which multiple audiences easily collapsed into one, jeopardizing individuals' privacy (see, e.g. Marwick & Boyd, 2011). For this very reason, Siibak and Tamme concluded, Estonian families favoured synchronous chat groups and other closed online spaces for their intra-family communication. This is an important observation, as, according to previous research, it is face-to-face conversation and telephone calls that have predominated as modes of family communication and in the maintenance of local relationships (e.g. Baym, 2015; Chen, Boase, & Wellman, 2002; Quan-Haase, Wellman, Witte, & Hampton, 2002).

All in all, compared to traditional person-to-person communication channels such as voice calls and short text messaging (SMS), instant messengers are particularly useful as tools helping people to stay in touch with closely related others and create and maintain communities based on closed communication spaces instead of public or semi-public social media platform use (Church & de Oliveira, 2014). Close-knit communities like families do not aim to reach large audiences, but are not limited to private one-to-one communication, either. Its ability to help users reach middle-range audiences, consisting of the significant others who all know one another, is thus one of WhatsApp's strengths.

Reaching the Family

The key informant reports in this study revealed marked country differences in the use of WhatsApp for intra-family communication. Some of these differences had to do with communication cultures and housing arrangements characterizing the context, such as a higher proportion of multigenerational households in Slovenia and a later home-leaving age in Italy, compared to Finland. The differences were, accordingly, directly related to physical distances between adult children, their parents and their grandparents, which were notably greater in Finland than in Italy and Slovenia. Physical distance from one another can thus be assumed to reinforce the need for electronically mediated family communication.

Another factor that can be presumed to either encourage or discourage the shift from voice calls and short text messaging to online-based communication is the prevailing pricing model for wireless Internet services. In Finland, mobile broadband subscriptions have typically included unlimited data transfer at a flat fee, while the rates in Italy and Slovenia are, as a rule, for limited service. Notwithstanding such differences, a common incentive for using WhatsApp in family communication in all the countries examined was cost saving. As several key informants reported, sending messages and making voice calls via WhatsApp were in their families considered free of charge, and hence a cheaper option compared to normal phone calls or conventional text/multimedia messaging that are often charged per-use (e.g. key informants Carla, aged 23, and Ella, aged 24, in Finland; Alice, aged 23, and Elisa, aged 26, in Italy).

At the time of this data collection, WhatsApp usage in Slovenia was largely confined to peer-to-peer communication: only one family of those surveyed in the country reported using it for intra-family communication. A couple of the Slovenian key informants, however, reported Viber, another instant messenger, as being used in their families, although mainly for the purposes of contacting distant relatives. Due to this, very limited role that WhatsApp played in the Slovenian families studied, the discussion below is confined to Finnish and Italian families.

In Italy, much of the family WhatsApp use reported took place among younger family members of approximately the same age. This communication, furthermore, was not restricted to closest family members only (e.g. key informants Alessandro, aged 20, Bruno, aged 27 and Matteo, aged 24), but was also resorted to reach cousins and second cousins (Silvia, aged 25) and, in some cases, also uncles and aunts, who, however, were normally less than 20 years older than the key informant (Melissa, aged 25, Monica, aged 25 and Enrico, aged 24). This pattern is in apparent agreement with the notion of the family that in Italy is broader than in Finland.

Families in which all family members used WhatsApp were clearly more common in Finland. Many Finnish key informants described WhatsApp exchanges the new daily mode of family communication in the families they reported on (e.g. those of Jenny, aged 25 and Sara, aged 25). This, to be sure, was still something of a new phenomenon in the families, as the family chat groups used for the purpose had been set up quite recently. One of the Finnish key informants reporting daily WhatsApp use in their families was Emma (aged 24), who spoke of the pivotal role a shared WhatsApp chat group played in her family's daily communication routines as follows:

> Me and my core family's [parents, aged 52 and 53, and sister, aged 19] main way of communicating is nowadays a WhatsApp chat group. We created this group about half a year ago, and it's come to very busy use ever since. One of us posts photos and messages for the group every day—and all the others follow them enthusiastically. The biggest difference with how we had it in the past is that now also my parents have learnt instant messaging on WhatsApp.

A major advantage of WhatsApp chat groups is that it allows reaching the entire or almost all of the family at once. The Finnish key informant Emilia (aged 24) made a point of noting this: 'Recently, we created a WhatsApp chat group for the family, so that we can easily reach all of us when we need to contact everybody at the same time.' Although the use of chat groups was not as common in Italy as it was in Finland, WhatsApp was clearly becoming more common in families there, too. Monica (aged 25) from Italy told that WhatsApp was something quite new to her family, and that it was mainly her mother (aged 53) and her siblings (aged 25 and 19) who used it:

> For instant messaging, we all use mostly the smartphone application WhatsApp. My dad [aged 58] is kind of a geek who likes technology. He tries to keep up to date with it but, because of his age and lack of time, he is not able to use WhatsApp as proficiently as the rest of us. To my mother [aged 53], although she's not the oldest of my respondents [family members], WhatsApp is a bit of a novelty as she got a smartphone only very recently. My aunts and uncles instead use it regularly, to chat with family and friends and to send photos to people.... I myself use WhatsApp with all my respondents, although especially with my mother, because I want her to learn how to use it and because I want to share parts of my life with her, since we live far away from each other.

The end of this quote illustrates well the sharing-as-caring aspect of WhatsApp-based family communication. Perhaps not so surprisingly, it was mostly visible in interactions between mothers and their daughters. This gendered aspect of WhatsApp communication was manifested in the reports of several Finnish key informants as well. Emma (aged 24), for instance, described how her mother (aged 52) long resisted the idea of acquiring a smartphone. When she finally received one from her employer and learnt to use it, it, however, quickly became her, the mother, who actively began putting it to collective use; it was then also she who 'came up with the idea of creating a WhatsApp chat group for the family'. Another Finnish key informant, Emilia (aged 24), captured the central role that mothers usually had in family WhatsApp communication, recounting how 'mom [aged 52] no longer needs to call her kids once a week to ask how they are doing, as now we exchange news every day'.

This pivotal role of mothers in family communication became all the more clearer when juxtaposed to fathers' more limited communication skills and practices. While the Finnish key informant Julia (aged 21), her sister (aged 19) and her mother (aged 54) all praised WhatsApp for being 'the best communication mean as it can be used for free to send messages all over the networks and, what's best, sending photos is so simple and costs nothing', they had nevertheless chosen another way to talk with the father in the family (aged 59). According to Julia, she, her sister and her mother 'always call [him] since he has not installed WhatsApp in his smartphones and his messages anyway are so messy and hard to read', as the father did not use punctuation in his messages, made lots of spelling mistakes and sent jokes the others did not understand. Also, some of the Italian key informants spoke of similar differences between mothers and fathers in WhatsApp communication. As Silvio (aged 21), for instance, reported:

> To keep in touch with my mom [aged 50] I can make phone calls or use texting, WhatsApp messages, or email, since she has been able to integrate herself almost completely into the world of technology, including using a smartphone. My dad [aged 54], on the other hand, is still at a lower step, so I only talk to him by phone, or I send him SMS's or, more recently, emails.

As these quotes clearly indicate, even when the entire family could be reached through WhatsApp, mothers were typically the main agents of family communication. Relatedly, there was frequently a fear that fathers would end up being left out if they did not learn or want to use instant messengers (e.g. by Finnish key informant Teresa, aged 24). In this sense, WhatsApp family communication, as it were, emerged as a new form of immaterial labour, in particular, care work, which still today remains more of a domain of women than men (cf. Fortunati, Taipale, & de Luca, 2013; Hochschild, 1983).

The importance of WhatsApp communication for the social coherence of family was clearly articulated by both Finnish and Italian informants: it facilitated intergenerational connections and togetherness within extended families. Sauli, the brother of the Finnish key informant Sofia (aged 24), testified to this: 'Thanks to WhatsApp, we write to and keep in touch with each other more often now'. Similarly, the key informant Emma (aged 24), also from Finland, reported that '[w]e spoke of how all of us had noticed how, after adopting WhatsApp, we have been much more often in touch with other family members than before.' Even a very young sister (aged 9) of the Finnish key informant Maria (aged 24) was able to notice the benefits of having a common WhatsApp chat group for her family: 'you know better how the other family members are doing, even when they are far away'.

In some families where the parents had not yet embraced WhatsApp, the children nevertheless felt that their doing so 'would make family communication easier', as a younger brother (aged 12) of the Finnish key informant Marika (aged 20) put it. A sister (aged 21) of the above Finnish key informant Sofia made the same point a little more concretely:

> my parents [aged 51 and 48] are excluded, so we have own small circle. The parents are a bit bitter because of this, as that they don't get to see the pictures we're sending each other…. Our relationships would be saved if only they, too, joined WhatsApp.

A Technology of Middle Reach

In her book *Personal Connections in the Digital Age,* Baym (2015) argues that the success of social networking sites is owing to their wide, but selective, reach. The notion of reach here is borrowed from Gurak (2003, p. 30), who describes it as 'the partner of speed': digitized contents not only travel with speed, but they can also reach large audiences. As Baym rightly notes, media technologies vary in their ability to attain, support or reach audiences of different sizes. The reach of face-to-face contacts is obviously the narrowest, while the qualities that in-person communication can mediate are by far most. In-person communication involves a range of non-verbal (facial and bodily) cues that are extremely difficult to mediate in full detail using technological means. Mobile media and communication technologies allow both a *narrow reach*, confined to one's closest friends and family members (when using phone calls, short text messages), and a *wide reach*, extending also to acquaintances and even strangers (through Twitter, Instagram, Facebook and so on).

However, as Austin (2017) has pointed out, Byam's observations on electronically mediated interpersonal relationships concern fairly early forms of ICT and social media. Instant messaging applications like WhatsApp that feature closed group chat functions seem to fall between the two extremes of narrow and wide reach. What they do, namely, is to enable one to create, access, sustain and manage a *middle-reach* audience. The extended family serves as a good example of such a middle-range community since it typically involves not only very close family members like siblings and parents, but also more distant family members and relatives such as step-parents and half-siblings, or grandparents living further away. Research relying on rather simple distinctions between weak and strong ties all too easily views today's families as loose nexuses of individually networked family members who merely need to make more efforts than those in the past to be able to stay connected (e.g. Rainie & Wellman, 2012).

In this chapter, my argument is that WhatsApp and similar mobile instant messengers have, in fact, introduced a whole new layer to mobile communication, one that helps to make this laborious task of families easier to accomplish. Family WhatsApp use does not simply bring together separate individual networks or conjoin family members who all know each other already. It also provides a relatively private communication space suitable for the sustenance and maintenance of both dyadic family relationships and entire family communities, allowing family members to discuss private family matters, exchange emotions and provide care and support to one another while keeping their exchanges and actions hidden from the larger public.

Here, the activity of sharing-as-caring attains then a deeper and fuller meaning. While minor acts of sharing, such as 'sharing' and 'liking' contents on Twitter, Instagram or Snapchat, might be sufficient to establish and maintain weak ties between users, strong ties are seldom, and family ties never, established purely online. Strong family ties require a great amount of time, emotion, intimacy and reciprocal services invested by family members in their intra-family relationships (Granovetter, 1973). What private family WhatsApp groups do is offer a particular channel to maintain and nurture strong family ties from afar and near, allowing both synchronous and asynchronous modes of communication that help family members juggle their individual daily agendas and timetables. Furthermore, considering that sharing, as an activity, in itself manifests values that are typically feminine (such as openness and mutuality; see, e.g. Johns, 2013), it is unsurprising that WhatsApp is used more widely by women than by men.

All in all, such affordances provided by closed WhatsApp chat groups resonate well with the particularities of contemporary extended families that are geographically dispersed, non-hierarchical and change their composition over time. Its new communicative properties have made WhatsApp and comparable applications extremely well suited for one-to-group type of communication, offering a platform for constant family connectivity (Hänninen, Taipale, & Korhonen, 2018; Ling & Lai, 2016). Thanks to them, family members who, to borrow the words of Rainie and Wellman (2012, p. 162) used to 'mostly dance solo but take part in a few duets and household ensembles' can now keep their own band together and play their joint favourite tunes non-stop if they so wish.

The larger meaning of sharing and exchanging small messages, photos and video clips, not forgetting nanolevel interaction such as pressings of 'like' and 'favourite' buttons (Eranti & Lonkila, 2015), is perhaps best captured by the concept of *phatic communion*. The term was used first by Malinowski (1923), who coined it to refer to apparently purposeless speech acts such as polite small talk and trivial pleasantries that nevertheless have an important social function in establishing, maintaining and renewing social bonds between interlocutors. As Miller (2008) has argued, online media cultures promote similar kind of, mainly social and networking driven, communication at the expense of functional and informational contents and dialogic intents. The design of many social media platforms, for instance, encourages short expression by limiting the number of characters that can be used for text input (e.g. Twitter), favouring the use of visual material and introducing new ways to graphically express emotions with one click. To critical voices such as Wittel's (2001), however, this has sounded like inviting a flattening of communication and even of social bonds.

In contrast to such more pessimistic predictions about the effects of the digitalization of also family communication on group cohesion and contacts, extended families in Finland, and increasingly in Italy as well, have discovered the positive potential of WhatsApp. For them, it has been a useful tool helping them to reach and keep in contact with their members, sustain family connection and maintain a sense of togetherness. Instead of making them stay only loosely connected and work even harder than before to keep in touch, family members have found multimodal communication and group chats via WhatsApp to facilitate intra-family communication and make it easier to reconnect with family members elsewhere. In other words, WhatsApp has helped in refreshing and reactivating social bonds between family members, and in so doing it has effectuated the transposition of the original function of phatic communion to the online environment.

Short, Fast and Trivial

Indeed, the role of WhatsApp as a medium of phatic expression was widely recognized in both Finnish and Italian families in this study. Most often, this recognition was indirect, expressing itself through a downplaying of the importance of the small messages one sent via WhatsApp, which nevertheless appeared to in many ways act as the basics of people's everyday family interaction. The report by the Finnish key informant Ella (aged 24) spoke of the suitability of the application for this purpose: 'at times, the contents of messages are not really important and full of information, and that is when WhatsApp is the best choice'. Similarly, a sister (aged 25) of the Finnish key informant Teresa (aged 24) opinioned that, in their family, 'WhatsApp has made us closer as we can speak about trivial matters and have fun even if we are physically in different places'

Another characteristic of phatic WhatsApp-mediated expression involved its adaptability to different kinds of communication needs and preferences. To facilitate intra-family communication and keep everybody in the family connected, family members often faced a need to accommodate everyone else's needs and preferences and adapt themselves to others' favourite communication modes. This they could more easily do with WhatsApp, as, among others, the Finnish key informant Karin (aged 27) and the Italian key informant Monica (aged 25), respectively, testified:

> With my partner [aged 23] and my little brother [aged 23], we communicate over our mobile phones, mainly by WhatsApp messages and through Facebook Messenger…. WhatsApp messages are usually the easiest and fastest ways to connect, if you want to talk to people belonging to a younger generation. Another major reason for why people use it is that it's free.

It should say that WhatsApp is something everybody agrees about, and in my opinion, it connects between different generations, as it allows the kind of short and fast communications my mother prefers, but also longer casual chats with lots of links and images, the kind of communication my sister likes.

As the Finnish Karin notes above, WhatsApp lends itself well to, and even promotes, short and quick communications favoured by younger people. Those communications can, furthermore, make use of emoticons and chat slang. At the same time, however, as the Italian Monica points out, unlike, say, Twitter with its 140-character limit, WhatsApp does not exclude any longer forms of expression, either. What Monica's account also reveals, however, is that our stereotypes regarding generation-specific communication styles do not need to always hold: among the families partaking in this study, there were situations that considerably differed from, or even reversed, them. In Monica's family, for instance, it was, in fact, her mother who favoured short, matter-of-fact-like exchanges, while the children enjoyed engaging in longer discussions.

Multimodality Spiced with Playfulness

A great deal of WhatsApp's popularity appears thus to be due to its ability to effectively and quickly transmit different types of contents. Sometimes, a seemingly purposeless exchange of photos and other media contents between family members may, in fact, provide the easiest way to engage in social bonding and share a sense of togetherness. In Finland, for instance, the key informant Carla (aged 23) told that, in her family, 'WhatsApp is what we choose, especially when we want to share photos with one another'. In many families, also parents had enthusiastically begun to exchange photos and videos to stay connected with the rest of the family, as in the following cases reported by the Italian key informants Antonio (aged 30) and Mario (aged 24), respectively:

Lately, especially during the holiday seasons, I have noticed how my parents [both aged 52] have begun to use WhatsApp more than before, although they still only use it for communications of minor importance or to share some photos and funny videos.

With my sister, my cousins who're my age, and my mother [aged 51] I sometimes also use the application WhatsApp, which is a very popular, convenient, and easy-to-use way to end videos and photos.

Such multimodality of family communication brings out some new aspects of it. First of all, the use of one's own voice and self-taken photos in messages makes communication more personal than what 'pure' texting is capable of achieving. The Italian key informant Alice (aged 23) described this effect when reporting that even though her 'parents [aged 55 and 56] didn't immediately understand the point with WhatsApp groups, they found in voice messages a new possibility to make their communications more personal compared to text messaging'. Second, the certain playfulness involved in the sharing of comical photos and videos for its part, too, contributes to the attainment of the ultimate outcome in all phatic expression, which is social binding. This could sometimes be reflected in what we could call the 'social division of labour' between different communication modes in the families, as had happened in the Italian key informant Francesco's (aged 25) case:

> I installed WhatsApp upon the request of my father [age N/A], who then began to flood my own smartphone with 'funny' videos he kept sending…. It's interesting to note that whereas I use WhatsApp for all communication, those in my father's and mother's [aged 57] generation tend to rather view it as more of a 'game', in the sense that they use it almost exclusively for unimportant things or to share entertainment. For everything else they do SMS and ordinary phone calls.

That parents, as in this quote, often try to engage many of the new communication modalities offered for users by WhatsApp and other applications like it suggests a willingness on their part to connect with their children more. Correspondingly, children's willingness to participate in family WhatsApp groups and their readiness to adapt their communication methods and styles to those of their parents speak of a similar desire to connect. When the audience consists of close persons of middle reach, such as those included in the closed family WhatsApp groups, intergenerational communication is quite immediate by nature and the risk of losing one' face is relatively low. In such a safe environment, it is then 'quite common to send greetings and funny videos, just so you can smile together', as Italian key informant Claudia (aged 21) summarized the purpose of using WhatsApp in her family.

Connecting Distributed Families

This chapter has examined WhatsApp as a technology of middle reach that serves the ends of social cohesion in extended families and intergenerational family relationships through its many modalities suited for phatic communion. In family WhatsApp communication, social bonding through small messages, endless everyday images and comical video clips are often more consequential or valued than any exchange of substantive information. In dyadic family relations, WhatsApp's many modalities allow family members to individually choose the method of communication most desired and suitable for each one of them. Perhaps even more importantly, however, WhatsApp provides a relatively safe environment for one-to-group communication. As a consequence, the informants in this study could report WhatsApp group chats to have clearly facilitated intra-family communication in their families and strengthened the cohesion of their geographically distributed extended families.

To conclude, WhatsApp appears to provide a well-functioning platform for facilitating intergenerational communication in families, especially between still young family members and their late-middle-aged parents. To the extent that parents, too, have begun using it to send photos, video clips and voice messages, it might, moreover, even be argued that WhatsApp is marking a shift away from any clear-cut distinction between 'texting teenagers' and their 'talking parents'. Especially, the mothers of the Finnish and Italian key informants in this study had embraced instant messaging as a means to communicate with their grown-up children, extending their role as family carers to the domain of electronically mediated communication and the Internet. What was also interesting to note, however, was that some parents, especially fathers in Finland, could also feel themselves excluded from online family communication (for more on this, see Hänninen et al., 2018). At the same time, it did, to be sure, also become obvious that instant messaging had not yet reached the oldest members of the extended families. Grandparents were hardly ever mentioned as active players in family instant messaging.

References

AudienceProject. (2016). *Insights: Social media and Apps in the Nordics*. Retrieved from https://www.audienceproject.com/wp-content/uploads/social_media_and_apps_nordics.pdf.

Austin, G. C. (2017). Personal connections in the digital age (2nd ed.). *Consumption Markets & Culture, 20*(3), 293–296.

Baron, N. S. (2010). *Always on: Language in an online and mobile world*. Oxford: Oxford University Press.

Baym, N. K. (2015). *Personal connections in the digital age* (2nd ed.). Cambridge: Polity.

Bouhnik, D., & Deshen, M. (2014). WhatsApp goes to school: Mobile instant messaging between teachers and students. *Journal of Information Technology Education: Research, 13,* 217–231.

Chen, W., Boase, J., & Wellman, B. (2002). The global villagers: Comparing internet users and uses around the world. In B. Wellman & C. Haythornthwaite (Eds.), *The internet in everyday life* (pp. 74–113). Malden, MA: Blackwell.

Church, K. & de Oliveira, R. (2014). What's up with WhatsApp? Comparing mobile instant messaging behaviors with traditional SMS. MobileHCI' 2013, August 27–30, 2013, Munich. http://dx.doi.org/10.1145/2493190.2493225.

Stern, C. (2015). Messaging will be Facebook's 'next major wave of innovation and financial windfall. *Business Insider Jun. 24, 2015.* Retrieved from http://www.businessinsider.com/facebooks-next-big-profit-driver-is-messaging-2015-6?r=US&IR=T&IR=T.

Eranti, V., & Lonkila, M. (2015). The social significance of the Facebook Like button. *First Monday, 20*(6). Retrieved https://doi.org/10.5210/fm.v20i6.5505.

Fortunati, L., & Taipale, S. (2017). Mobile communication: Media effects. In P. Rössler, C. A. Hoffner, & L. van Zoonen (Eds.), *International encyclopedia of media effects* (pp. 1241–1252). Hoboken, NJ: Wiley-Blackwell.

Fortunati, L., Taipale, S., & de Luca, F. (2013). What happened to body-to-body sociability? *Social Science Research, 42*(3), 893–905.

Granovetter, M. S. (1973). The strength of weak ties. *American Journal of Sociology, 78*(6), 1360–1380.

Grinter, R. E., & Palen, L. (2002). Instant messaging in teen life. Proceedings of the 2002 ACM conference on Computer supported cooperative work (pp. 21–30). New York, NY: ACM.

Gurak, L. J. (2003). *Cyberliteracy: Navigating the Internet with awareness*. New Haven, CT: Yale University Press.

Hänninen, R., Taipale, S., & Korhonen, A. (2018). Refamilisation in the broadband society. The effects of ICTs on family solidarity in Finland. *Journal of Family Studies.* Advance online publication. https://doi.org/10.1080/13229400.2018.1515101.

Hochschild, A. (1983). *The managed heart: Commercialization of human feeling*. Berkeley, CA: University of California Press.

Horst, H. A. (2006). The blessings and burdens of communication: Cell phones in Jamaican transnational social fields. *Global Networks, 6*(2), 143–159.

Johns, N. A. (2013). The social logics of sharing. *The Communication Review, 16*(3), 113–131.

Johnston, M. J., King, D., Arora, S., Behar, N., Athanasiou, T., Sevdalis, N., et al. (2015). Smartphones let surgeons know WhatsApp: An analysis of communication in emergency surgical teams. *The American Journal of Surgery, 209*(1), 45–51.

Karapanos, E., Teixeira, P., & Gouveia, R. (2016). Need fulfillment and experiences on social media: A case on Facebook and WhatsApp. *Computers in Human Behavior, 55*, part B, 888–897.

Lenhart, A., Rainie, L., & Lewis, O. (2001). *Teenage life online: The rise of the instant-message generation and the Internet's impact on friendships and family relationships.* Washington, DC: Pew Internet & American Life Project. Retrieved from http://www.pewinternet.org/wp-content/uploads/sites/9/media/Files/Reports/2001/PIP_Teens_Report.pdf.pdf.

Lim, S. S. (Ed.). (2016). *Mobile communication and the family.* Dordrecht: Springer.

Ling, R. (2007). Children, youth, and mobile communication. *Journal of Children and Media, 1*(1), 60–67.

Ling, R., & Baron, N. S. (2007). Text messaging and IM: Linguistic comparison of American college data. *Journal of Language and Social Psychology, 26*(3), 291–298.

Ling, R., & Lai, C. H. (2016). Microcoordination 2.0: social coordination in the age of smartphones and messaging apps. *Journal of Communication, 66*(5), 834–856.

Madianou, M., & Miller, D. (2011). Mobile phone parenting: Reconfiguring relationships between Filipina migrant mothers and their left-behind children. *New Media & Society, 13*(3), 457–470.

Malinowski, B. (1923/1994). The problem of meaning in primitive languages. In C.K. Ogden & I.A. Richards (Eds.), *The meaning of meaning* (pp. 435–496). London: Routledge.

Marwick, A. E., & Boyd, D. (2011). I tweet honestly, I tweet passionately: Twitter users, context collapse, and the imagined audience. *New Media & Society, 13*(1), 114–133.

Miller, V. (2008). New media, networking and phatic culture. *Convergence, 14*(4), 387–400.

Nag, W., Ling, R., & Jakobsen, M. H. (2016). Keep out! Join in! Cross-generation communication on the mobile internet in Norway. *Journal of Children and Media, 10*(4), 411–425.

Nardi, B. A., Whittaker, S., & Bradner, E. (2000). Interaction and outeraction: instant messaging in action. In Proceedings of the 2000 ACM conference on Computer supported cooperative work (pp. 79–88). New York, NY: ACM.

O'Hara, K., Massimi, M., Harpe, R., Rubens, S., & Morris, J. (2014). *Everyday dwelling with WhatsApp.* In Proceedings of the 17th ACM conference on Computer supported cooperative work & social computing (pp. 1131–1143). New York, NY: ACM.

Quan-Haase, A., Wellman, B., Witte, J. C., & Hampton, K. N. (2002). Capitalizing on the net: Social contact, civic engagement, and sense of community. In B. Wellman & C. Haythornthwaite (Eds.), *The internet in everyday life* (pp. 291–324). Oxford: Blackwell.

Rainie, L., & Wellman, B. (2012). *Networked: The new social operating system.* Cambridge, MA: MIT Press.

Ribak, R. (2009). Remote control, umbilical cord and beyond: The mobile phone as a transitional object. *British Journal of Developmental Psychology, 27*(1), 183–196.

Rosales, A., & Fernández-Ardèvol, M. (2016). Beyond WhatsApp: Older people and smartphones. *Revista Română de Comunicare și Relații Publice, 18*(1), 27–47.

Siibak, A., & Tamme, V. (2013). 'Who introduced granny to Facebook?': An exploration of everyday family interactions in web-based communication environments. *Northern lights: Film & media studies yearbook, 11*(1), 71–89.

Statista. (2017). *Number of monthly active WhatsApp users worldwide from April 2013 to December 2017.* Retrieved from https://www.statista.com/statistics/260819/number-of-monthly-active-whatsapp-users/.

Taipale, S., & Farinosi, M. (2018). The Big Meaning of Small Messages: The Use of WhatsApp in Intergenerational Family Communication. In J. Zhou and G. Salvendy (Eds.): *Human Aspects of IT for the Aged Population 2018, Lecture Notes in Computer Science* (pp. 532–546). Cham: Springer.

Taipale, S., Petrovčič, A., & Dolničar, V. (2018). Intergenerational solidarity and ICT usage: Empirical insights from Finnish and Slovenian families In S. Taipale, T.-A.- Wilska, & C. Gilleard (Eds.), *Digital technologies and generational identity: ICT usage across the life course* (pp. 68–86). London & New York, NY: Routledge.

Taloustutkimus. (2014). *Social media survey for YLE*. Retrieved from http://www.yle.fi/tvuutiset/ uutiset/upics/liitetiedostot/yle_somekysely.pdf.

Wilding, R. (2006). 'Virtual' intimacies? Families communicating across transnational contexts. *Global Networks, 6*(2), 125–142.

Wittel, A. (2001). Toward a network sociality. *Theory, Culture & Society, 18*(6), 51–76.

Chapter 8
Intergenerational Solidarity

Abstract This chapter considers whether and how changes in the maintenance of digital home and familial relationships might be linked to the ways in which intergenerational family solidarity is expressed in extended families. Drawing upon Bengtson and Roberts's intergenerational solidarity model, the argument is made that new media and communication technologies are associated, in particular, with associational and functional forms of intergenerational solidarity, while normative, affectual and consensual forms of solidarity are expressed to a far more limited degree. Country differences in the manner and extent to which new technologies and applications are used for intra-family communication are identified and discussed, as is the impact of personal health conditions on the employment of these technologies for family communication.

Keywords Digital technologies · Family norms · Functional help · Health condition · Intergenerational solidarity · Living arrangements

Having, in the previous three chapters, developed a basic understanding of the changing family roles, responsibilities and practices in the maintenance of digital home and familial relationships, we may now look at whether and how the changes involved might be linked to the ways in which intergenerational family solidarity is expressed in extended families. Drawing upon the intergenerational solidarity model developed by Bengtson and Roberts (1991) and introduced in Chap. 4, the argument will be made that new media and communication technologies are associated, in particular, with associational and functional forms of solidarity between generations, while normative, affectual and consensual forms of solidarity are expressed to a clearly more limited degree. Any country differences found in the manner and extent to which new technologies and applications are used for intra-family communication are discussed, as is the impact of personal health conditions on the employment of these technologies for family communication.

An earlier version of this chapter was originally published as Taipale, Petrovčič, and Dolničar (2018).

Social Bonding

The most significant single message emerging from the key informants' reports was that new media and communication technologies facilitated intra-family communication, especially in Finland. A general understanding appeared to be that social media applications, in particular, such as WhatsApp, Facebook, Path and Instagram, had not only increased, but also enriched the interactions among family members. In Finland, it was the possibility to use, besides voice calls and text messaging, also photos, videos and voice messages that was considered as especially enriching in this regard (e.g. by the key informants Teresa, aged 24, and Eva, aged 23). Most often, however, the potential positive effects of WhatsApp a new platform for family communication were seen as relating to improved intergenerational family connections (see also Chap. 7). In Finland, Sofia's (aged 24) brother Johan (aged 21) testified to such benefits as follows: 'Thanks to WhatsApp, we write to and in general communicate with one another more than before'. Similarly, also another Finnish key informant, Emilia (aged 24), maintained that 'after adopting WhatsApp, all of us are much more in touch with other family members'. WhatsApp, for these families, had not only made interactions more regular but had also increased the volume of communications and enabled one-to-many communication. As the Finnish Emma (aged 24) described these changes:

> They [other family members] think that our WhatsApp group and the fact that we use it regularly have brought us closer to one another. Now, with everybody receiving the same messages at exactly the same time, we can talk to the entire family, not just one to one. This is especially important for our dad, since we would otherwise call our mom more often; now our communication within the family is more balanced.

This quote reveals another characteristic of associational solidarity in Finland. There, especially grandparents, who did not use the same communication technologies as their younger family members, but often also middle-aged fathers were either not included or themselves opted not to participate in other family members ICT-mediated communication networks (as, e.g. in the family of Isabella, aged 22). As Sara (aged 25), for instance, described the situation in her family:

> Most of the messages I send [via WhatsApp and Facebook] are to my mom [aged 54], and the same applies to my brother [aged 33], too. Sometimes I send stuff to my brother, too, but often I prefer to call instead, if it's him. To my dad [aged 59] we seldom send any messages via WhatsApp, as his Internet connection is not always on, and so he does not notice the messages when they arrive.

Also, other reports from Finland spoke of limited communication with fathers. Instead of talking to him directly and asking, a 22-year-old sister of Isabella (aged 22), for example, explained that she would rather 'count on my mother [aged 49] to convey any news from my father [aged 52]'. In some families, fathers had developed a distinct sense that they had been excluded from all the talking and chatting, expressing it to others. As, for instance, the Finnish key informant Julia (aged 21) told, '[m]e and my sister [aged 19] have noticed that our dad [aged 59] thinks that we constantly chat on the phone with our mom [aged 55] only, without calling him nearly as often. Which is indeed partly true'.

These excerpts appear to confirm the understanding that mothers have remained the main agents of family communication, at least in Finland. In some of the Finnish families, it was also the mother who had initially suggested creating a WhatsApp chat group for the family's purposes (e.g. in the family of Emma, aged 24). As one of the Finns, Teresa (aged 24) summarized the dominant role of the mother (aged 58) in her family, compared to the father, 'WhatsApp is used in our family by our mother and all the children'. Besides WhatsApp, mothers were connected to their children through other social media such as Facebook and Instagram (e.g. the family of the Finnish Maria, aged 24).

In Italy, the use of new mobile messaging applications was concentrated in younger family members keen to find new ways to stay connected. The report by Anita (aged 28) was illustrative in this respect: 'Young family members instead utilize their smartphones broadly, using many messaging applications and different social networks to stay in contact with people of the same age or younger' (the same was described the Italian key informant Bruno, aged 27). Besides instant messaging, also mobile phones were commonly deployed for daily key informant–parent communications in the country. According to Antonio (aged 30), these interactions involved 'primarily the mobile phone, with the communication being in most cases through voice-only calls, and to a much lesser extent text messages' (also, e.g. the Italian key informants Martina, aged 21, and Enrico, aged 24). Just as in Finland, also some of the Italian key informants, like Monica (aged 25), drew attention to the particularities of the mother–children relationship. What prompted Monica to make good use of the associational capacity of WhatsApp was the strong bond she had with her mother, combined with the distance that physically separated them.

Contrasting with Italy, in Slovenia the country's common multi-generational living arrangements and short geographical distances did not in the same way lead people to start using digital technologies for family communication. The Slovenian key informant Katarina's (aged 26) case may serve as an example:

> I communicate with other members of my family via my mobile phone, which is enough because we live so close to and regularly visit one another, so there's no need for us to use Skype, Facebook, and so on.

Occasionally, when a family member travelled abroad or went to visit a relative far away, Skype and social media were used to maintain contacts, and this was so in all the three countries (e.g. the families of Mia, aged 25, and Anton, aged 29, in Slovenia; the families of Sara, aged 25, Lucas, aged 38, Maria, aged 24, and Emma, aged 24, in Finland; the families of Melissa, aged 25, Marco, aged 24, and Martina, aged 21, in Italy). The same geographical separation factor made also older family members realize the value of the associational capacity of new technologies, as the following quotes from the reports by the Slovenian key informant Katarina (aged 26) and the Italian key informant Melissa (aged 25) demonstrate:

> Since my sister [aged 29] lives [800 km away] in Rome, the family members talk to her via the Internet—the Skype. In the last six years, the grandfather [aged 82], too, has been learning how to use Skype, since he wants to communicate with his granddaughter who comes and visits Slovenia only twice a year. (Katarina)
>
> I noticed, in fact, that I use instant messaging applications only with my aunt and my dad, who own a smartphone, and with the latter only when I am abroad, since that way we can stay in touch for free. (Melissa)

To summarize, country differences in family structure and living arrangements played a major role in influencing the degree and extent to which new digital technologies were used for intra-family social bonding in the three countries studied. While the geographically scattered extended families in Finland had found the exchange of short messages via WhatsApp and Facebook to provide the best way for them to perform family solidarity from afar, the physical proximity of family members in Slovenia did not entail a similar role for technologies in family communication. Italian families fell in between these two extremes: new personal communication and social media tools were used to stay in touch with family members, but this was so mainly among the younger ones among them since, at the time of the data collection (2015), it was only recently that their parents had begun to explore and adopt them.

Affectual Relations at Stake

In the research material, there was not much evidence of ICTs somehow particularly contributing to the exchange of either positive or negative sentiments. However, what the evidence did reveal was that good affectual relationships between children, parents and grandparents facilitated the uptake of new technologies in all of the countries. Intergenerational reassurance, for instance, was considered in many families as a factor promoting ICT usage among older family members. It allowed other family members to be encouraged 'to try to find a solution on their own' when hands-on teaching in connection with the adoption or use of an ICT was not enough, as Marija (aged 25) from Slovenia described it. Another Slovenian key informant, Petra (aged 25), confirmed the usefulness of this state of affairs, stating as her own observation, too, that '[t]hey [older family members] first need some encouragement'.

Even though the need to provide practical and affectual support to others was also felt to be a burden sometimes, younger family members at the same time appreciated being considered as useful to those others. In Slovenia, for instance, this was the case with Jakob (aged 26): 'I would probably say that those of us who help the others in such moments feel good about themselves because of it—because we feel useful and we are happy to help'. Also, in Finland, younger people tried to help their older relatives by advising them in the use of digital technologies. Simon (aged 24), for instance, reported that 'I have noticed that I often assume the role of someone encouraging others in technology use. I'm happy to give advice, though, and I try to motivate for instance my grandmother to use Skype more'.

Reversely, the exclusion of some family members from ICT-mediated family communication could stem from a lack of affectual solidarity in the family. This appeared to often be the case especially in Finland, where grown-up children frequently thought of their older relatives' postings on Facebook or Instagram as 'embarrassing' (e.g. the family of Maria, aged 24). In addition, there could be situations descriptive of this state of affairs like that of the Finnish key informant Rita (aged 34), who reported that 'phone calls with my father tend to be uncomfortable, so I rather send him text messages or talk face-to-face'. The Italian Emma (aged 24) in turn estimated that, for older people in her family, 'the communication taking place on social networks or through instant messaging is often unnecessary, so they prefer, instead, real, "live" exchanges that involve more emotions'. Another Italian key informant, Alessandro (aged 20), wrote along the same lines that:

In my opinion, for my grandparents, who are rather old already, it is easier, if not more natural, to show affection towards relatives in general. For younger people, let's say between cousins or brothers, it is more difficult at an affective level to show emotions, as these days you show affection simply by, for instance, sending a poke on Facebook, or some emoticons on WhatsApp, or photos via Snapchat.

The other side of the coin, however, was that 'in some more complicated relations, social media is a low-threshold medium for expressing warm emotions difficult to show face-to-face or difficult express in words', as another Finn, Laura (aged 29) maintained in her report. These contradictory examples illustrate well how social media and other digital technologies are used with careful consideration to address the varied needs of families. Digital technologies can, however, only enhance intergenerational affectual solidarity in families if everyone in them uses the same tools and applications.

In Search of Consensus

Consensual solidarity refers to the degree of agreement or disagreement with beliefs, values or life orientations related to ICT use for family communication. What needs to be noted here is that this study did not address itself to any such shared values or beliefs; yet, some such were clearly taking shape in families where parents followed their children's ICT use. Sometimes, however, the ICT skills gap between generations was regarded as a barrier for any formation of consensual solidarity. As the Slovenian key informant Erik (aged 25) described one such situation:

He [father] was not always so confident and technologically fluent, but with my help and seeing my enthusiasm for new technologies, he has become a bit of a connoisseur of ICTs, although he still doesn't himself fully understand the full range of his knowledge, which can in fact rival mine in some areas.

There was, in general, a growing agreement in the families about how important developing one's ICT skills were (the families of, e.g. the Slovenian Marija, aged 25, Tia, aged 26, and Anton, aged 29). This consensus was built upon the idea that not everyone needed to have the exact same skills; skills that complemented each other were more important, making as they would the family as a whole stronger. The Slovenian Marija (aged 25) spoke of this as follows:

> They [father, mother, aunt, in their 60s and 70s] all stress the importance of communication skills, which are very important in everyday life, in interpersonal interactions, and in ICT use. They do not compare any skills directly with one another, as they see them to represent different categories of skills that are not comparable to one another; instead, they are intertwined and, as they say, all very important for the successful and well-integrated functioning of our everyday life and our ICT tools.

The most obvious disagreement, in particular, in Finland, related to the question of what constituted proper online communication. While the younger family members were more accustomed to open and straightforward online communication, the older ones called for thoughtfulness and linguistic correctness. This difference between family generations was addressed explicitly by Rita (aged 34) in Finland, who told that:

> My parents [aged 66 and 72] are horrified about all that openness that my sister [aged 26] demonstrates on Facebook and in her own blog. But my sister says that she hasn't noticed that her openness would have caused her any harm, in terms of her social relationships or her chances of finding a job, for example.

Also, the Finnish key informant Maria (aged 24) wrote of such differing conceptions: 'My father [aged 50] speaks of how parents associate communicating to others with some degree of formal style, saying that everything should to be taken quite seriously, that everything you say should be considered carefully. Among the younger people interactions are more easy-going and free'. The Slovenian key informant Veronika (aged 27) also wrote about such disagreements between generations regarding the proper style of communication:

> Those in the younger generation find it a bit strange and slightly disturbing when older people write their text messages in proper and a bit formal Slovene. Most of the young people are accustomed to messaging in simple, casual language.

Such a lack of shared values and orientation related to ICT use for family communication was indeed reported for some families in this study. The Italian key informant Alice (aged 23) described one such case indirectly, by telling how 'in my mom's [aged 54] opinion, it is really terrible when there isn't any collaborative spirit in our family' in this regard. Despite any disagreements along these lines, however, the research material overall spoke in the first place of families' efforts to reach consensus and overcome any generational stereotypes. For instance, the stepfather (aged 55) of Laura (aged 29) in Finland brushed of the lasting significance of any disagreements in this respect by noting how '[t]here are jerks in every generation' and speaking of 'how it is easier to avoid them [conflicts] in social media by leaving the scene'. Such consensus-favouring orientation was evident in the reports collected from all three countries. Consensus was also associated with the idea of the democratic family, a family in which everyone has an important, yet different role to play. Even if all informants in this study did not consider their family to be democratic in that way, family members seemed to generally agree that their roles had indeed somewhat changed (e.g. the family of Marija, aged 25, in Slovenia, see also Chap. 5). Even where parents still had the final word in technology-related family decisions, children were typically almost always listened to and consulted.

Functional Help in Technology Use

As concerns functional solidarity, two major themes emerged from the research material relating to how new media and communication technologies were used in digital families. Both of these have already been discussed above, in connection with the concepts of warm expert and life course.

The first theme was about equity in the exchange of knowledge and resources over the life course. When children are young, parents teach them some basic digital skills, as had happened, for example, to Marija (aged 25) in Slovenia, too: 'When I first started using my bank's electronic banking service a couple of years ago, my father was already using it—he had learnt it from my brother—and so he helped me to learn how to do it, too, which I greatly appreciated'. In Italy, the key informant Monica (aged 25) alluded to the same phenomenon as follows: 'Before, it was my father who would teach me how to use certain technology; now I teach him'. Sometimes parents and grandparents, however, also taught their adult children, such as in the case of domestic technologies not ordinarily needed when young. The Slovenian Veronika (aged 27) mentioned an electronic blood glucose monitor and a digital meat temperature gauge as examples of such technologies. It was, furthermore, largely agreed that parents could, in general, deepen their children's understanding of various issues, thanks to their life experience (e.g. Karin, aged 27, in Finland; sister of Natalija, aged 30, in Slovenia). Parents could also teach patience in ICT use (e.g. the families of Sara, aged 25, Mary, aged 26, and Simon, aged 24, in Finland) and help their children to understand differences between formal and informal styles of communication (as in the case of the Finnish Mary, aged 26). They, moreover, sometimes also talked to their children about online risks and advised in safe ICT use (as in the case of, e.g. Karin, aged 27, in Finland; niece of Tia, aged 26, in Slovenia). As children grew up, however, the teaching roles were typically reversed, especially when it came to learning about the more technical side of things.

The second major theme here pertained to the intergenerational provision of help by grandchildren. In Italian families, in particular, it was widely noted how, when older people in the family encountered difficulties with ICTs, younger family members supported them. This provision of help was, however, not described as regular or very intensive in nature; help was simply given when a need for it arose (e.g. the families of Alice, aged 23, Silvia, aged 25, and Bruno, aged 27). The situation in Italy was very much like that in Finland, where any interaction between grandchildren and their grandparents was typically limited to short telephone calls and text-message greetings only (e.g. the families of Sara, aged 25, and Emma, aged 24). In Slovenia, the grandchildren had a more close-knit relationship with their grandparents, which, among other things, often meant more regular provision of assistance in ICT use (e.g. the families of Franc, aged 25, Veronika, aged 27, Mia, aged 25, Tina, aged 25, and Katja, aged 25). There, the key informant Katarina (aged 26) describes one such relationship as follows, revealing also some of the demands it made on younger family members:

> It is an on-going process, so almost every Sunday when I visit him [grandfather, aged 83], I have to help him with something. I am also bothered by the fact that I often don't know what he needs help with…. For example, if it's about how to use Outlook, I, as a Gmail user, am not always able to understand what it is that he wants.

While young people in both Finland and in Italy helped their parents with sorting out various technical problems, they were less frequently in touch with their grandparents and were less intensively engaged with them than their Slovenian counterparts. Much of this country difference is explained by the greater geographical distances between children, their parents and grandparents especially in Finland, compared to Slovenia. Providing assistance in technological matters from afar was deemed as often very challenging, in particular, when the people in need of help were technologically clearly less savvy.

Unshaped Normative Solidarity

In the data collected by the key informants, there was scant evidence of the existence of any family norms concerning ICT use. Nevertheless, on certain specific issues like data security family members in the three countries tended to find themselves in even a spontaneous agreement with one another. As the brother (aged 21) of the Finnish key informant Sofia (aged 24) put it, 'When my parents tell me "don't download this and don't download that, even if you think you need them [for the functioning of a programme or the like]", I mostly do as they say'. Somewhat similarly, the Slovenian Katarina (aged 26) told about how her father (aged 58) 'is very reticent to publish any personal posts, and he keeps telling and teaching others not to publish any personal information on the web'. As Katarina noted, her 25-year-old sister 'is very aware of this, and she only posts things of a more general character'. However, it was only the two sisters who ended up acting based on this understanding, or norm, in the family, since the others did 'not post a lot of information online, focusing more on information searches'. In general, however, there was little evidence of how widely such and similar norms were acknowledged, shared and/or complied with in the families, as there was so much variation in ICT use both between and within families.

In all three countries, there were certain responsibilities that were entrusted to, and typically also accepted by, the persons acting as the warm expert in their family (see Chap. 5), to whom others would then turn for help (e.g. the families of Isabella, aged 22, and Carla, aged 23, in Finland; the families of Franc, aged 25, Tia, aged

27, and Veronika, aged 27, in Slovenia). The following quote by the Finnish key informant Karin (aged 27) illustrates this rather common situation:

> My brother has the main responsibility to ensure that our communication tools, applications, and programmes all work well. I feel like it is self-evident that he sorts out the problems if I detect any with our devices and programmes. I never hesitate to ask him for help, either.

What distinguished the Slovenian cases from the Finnish and Italian ones was the fact that the help the Slovenian grandchildren on a regular basis extended to their grandparents was seen as a kind of filial duty or a cultural norm, rather than ad hoc volunteering based on a case-by-case consideration. The Slovenian interviewees frequently took this provision of ICT assistance by grandchildren for granted, underscoring the normative nature of the expectations for it. Indeed, giving such aid was considered a natural part of family life, as expressed, for instance, for her part by the Slovenian key informant Petra (aged 25): 'Whenever my family members need any help, I am glad to assist them no matter how busy I might be. I feel it as a kind of duty to do so, because that is how I was raised'. Also, another key informant in the country, Anja (aged 21), wrote that her grandfather often preferred contacting his grandchildren directly when needing assistance, as they knew how to help.

What emerged from the Italian and Finnish data was certain ambivalence about the norms for how to communicate across generations. According to the Italian Antonio (aged 30), for instance, all older family people in his family had agreed that there was among younger people an 'obsessive necessity to check statuses, notifications, SMS, missed calls', even though they could also 'live the way we [older family members] did when we were young'. In Finland, Rita (aged 34) referred to certain normative expectations that governed communication differently depending on whether the question was of younger or older family members:

> My parents are...more dutiful and reliable as communicators than people of my own age or younger. They always answer the phone, if they are not driving a car or sitting in the sauna. They also reply to all text messages that they get, and they read them immediately when they hear that they have received one. They also answer to emails right away when they read them. For people my own age, the mobile phone etiquette is far more different from that with the fixed phones.... There is, for example, no need to always answer the phone, and you may even switch off your phone completely if you want to be alone.

As the quote shows, the normative basis governing the use of ICTs for communication had not been established in Rita's family yet. In general, disagreements between generations about the proper way of using ICTs (also, Colombo, Aroldi, & Carlo, 2018) appear to echo more general normative expectations that separate younger people from their elders. The same way, a relatively strong expectation concerning the provision of assistance from grandchildren to grandparents (or the absence of such expectations, as in Finland) cannot be considered specific to the use of ICT only: it, too, reflects more profound cultural values and the nature of the prevailing living arrangements, all of which vary between Scandinavian and South European countries (e.g. Hank, 2007).

Enabling and Preventing Factors

Structural solidarity refers to the opportunities for, or barriers to, intergenerational family interaction via new media and communication technologies. These structural factors shed light on many country differences discussed earlier in this chapter. While shorter geographical distances enabled regular in-person interaction between family members in Slovenia, longer distances in Finland created a demand for technology-mediated family communication from afar. As the Slovenian key informant Tia (aged 26) noted (also, e.g. Jakob, aged 26, and Angela, aged 27, in the same country):

> It's true that in our family we have always spent a lot of time together, and that we all live relatively close to one another, so, the time it takes to call someone, you might just as well go find them and tell them the same thing in person.

Contrasting with their Slovenian counterparts, Finnish interviewees frequently stressed that, regardless of any differing technology preferences between younger and older generations, their families were highly dependent on digital communication technology owing to long distances separating family members. The key informant Emma (aged 24) summarized the issue as follows: 'As there are several hundreds of kilometres between us, meeting face to face is not very often possible'. As she then concluded, however, 'information technology makes it possible for us to nevertheless maintain close relationships with our dear ones, even across great distances'. Also, Italian respondents brought up long distances as a reason for their intra-family mobile phone use, although in a considerably less pronounced manner (e.g. Monica, aged 25). In Slovenia, the importance of digital communication technology (email, Skype, Viber, etc.) for family communication and solidarity was recognized especially in situations where a family member moved to abroad, to another country (e.g. the families of Erik, aged 25, Julija, aged 25, and Klara, aged 28, in Slovenia).

Individual family members' personal health status and functional capabilities were other structural factors influencing the possibility to enhance family solidarity via ICT use. Poor eyesight and compromised hand agility were mentioned as factors reducing ICT use for family communication, especially in older family members. As the Finnish Emilia (aged 24), for instance, reported, '[m]y grandfather's [aged 85] vision has deteriorated so much that now he can barely read or write. He has also forgotten how his equipment works, so he no longer uses any other devices apart from his phone'. In Slovenia, Petra (aged 25), for her part, told of her parents and her grandmother who all used a feature phone while complaining that 'their fingers are too stiff and do not have much sensitivity left in the fingertips' (also Marija, aged 25, and Aleksej, aged 25, in the same country). Sometimes, however, the older family members had health problems that made one suddenly realize at least the potential benefits of ICTs. In the Slovenian Katarina's (aged 26) family, there had been one such case: 'My grandmother [aged 80] did not want a mobile phone, but then she got one anyway when she had to spend a long time in the hospital because of knee surgery'. Both of Katarina's grandparents (aged 80 and 83) were soon 'convinced of ICTs as a path to easy communication with their granddaughters and great-grandson'.

Solidarity on Display

This chapter has explored the way the use of digital media and communication technologies in extended families relates to family solidarity in Finland, Italy and Slovenia. The suggestion was made that, in this study, media and technology use fostered solidarity in extended families, although principally associational and functional solidarity only. The specific forms of solidarity found in families, and the strength of that solidarity, appeared to be dependent on the living and housing arrangements (whether or not several generations lived together in the same household) and life stage. Interestingly, the strength of the associational form of solidarity appeared to be inversely related to the physical proximity of other family members. The closer the family members resided to one another, the less there was need for using communication technology to strengthen the family bonds. As noted above, to be sure, also normative, consensual and affectual forms of solidarity were displayed, but to a lesser degree. One possible explanation for why this should have been so is the increased amount of individual networking in all the countries and families in this study. Individual networking activities require and attract less familial regulation in terms of how the digital media and communication tools ought to be used, and may thus rather cause disagreements within the family than add to its integration.

Some obvious country differences were detected in the form of intergenerational family solidarity that took the centre stage. In Finland, continuous exchange of short messages via mobile phones and social media applications promoted associational solidarity between family members who lived far apart from one another and thus had few occasions to meet in person. In contrast, the adoption and use of ICTs in Slovenian families fed functional solidarity between generations. Physical closeness in intergenerational relationships made helping one's elders in technology usage—a new form of social support between grandchildren and grandparents—natural and common as a family practice, with positive consequences for family solidarity. Finally, in Italian families, functional solidarity was perhaps not as prominently featured, widespread and expansive as in Slovenia, but it still involved a wider circle of family members (e.g. aunts and uncles) than in Finland, where family connections, in general, were more limited and distant.

References

Bengtson, V. L., & Roberts, R. E. (1991). Intergenerational solidarity in aging families: An example of formal theory construction. *Journal of Marriage and the Family, 53*(4), 856–870.

Colombo, F., Aroldi, P., & Carlo, S. (2018). "I use It correctly!": The use of ICTs among Italian grandmothers in a generational perspective. *Human Technology, 14*(3), 343–365. https://doi.org/10.17011/ht/urn.201811224837.

Hank, H. (2007). Proximity and contacts between older parents and their children: A European comparison. *Journal of Marriage and Family, 69*(1), 157–173.

Taipale, S., Petrovčič, A., & Dolničar, V. (2018). Intergenerational solidarity and ICT usage: Empirical insights from Finnish and Slovenian families. In S. Taipale, T.-A. Wilska, & C. Gilleard (Eds.), *Digital technologies and generational identity: ICT usage across the life course* (pp. 68–86). London and New York: Routledge.

Part III
Conclusions and Implications

Drawing upon grassroots-level interview data, in this part of the book I want to examine more closely the ways in which, and the extent to which, distributed families actually stay connected with the help of digital media and communication technologies and intergenerational cooperation needed to sustain the functionality of increasingly digital homes. Theoretical and conceptual implications of these dimensions of modern family life are discussed. As one of the main findings from this study, the argument is put forth that the changes we are witnessing in family roles, household tasks and responsibilities, as well as communication practices, do not simply or in some straightforward manner just erode family solidarity and the sense of unity within the family: they, at the same time, also provide new avenues for weaving family members together. In Chap. 9, the notion of re-familization is introduced, to allow for a better grasp of the cohesive impact of digital technologies in the context of extended and geographically distributed families. Chapter 10 then picks up from where Chap. 9 ends, delineating a more balanced approach to the study of digital families. It is claimed, among other things, that the kind of caring relationships that are performed and expressed via and in connection with digital technologies in the digital family can only be captured if the attention is turned away from individualized practices of technology use, towards the ways in which digital technologies are used within and for the family. The chapter concludes with a rough attempt to outline the future of the digital family.

Chapter 9
Technologies of Re-familization

Abstract In this chapter, the notion of re-familization is introduced, to allow for a better grasp of the cohesive impact of digital technologies in the context of extended and geographically distributed families. In the field of social policy, the notion of re-familization implies a reversal of the politics of de-familization that once was the hallmark of the golden-era welfare state. The argument is made that family-initiated uses of digital media and communication technology in response to (older) family members' daily help and care needs resonate well with the idea behind re-familization. In conclusion, the chapter presents several ways in which re-familization manifests itself in the everyday life of digital families.

Keywords Care and help needs · De-familization · Digital media · ICT ·
Re-familization · Social policy

While the concept of re-familization is not entirely new as such, it is new to media and communication studies. We will therefore do well to first take a quick look at where it comes from, before considering how it might be able to be useful for the study of digital families. Building upon the empirical investigations reported on in Part II, the chapter then goes on to propose that current digital media and communication technologies, which increasingly serve not just information-seeking needs, but also social and group communication needs, lend themselves well to being examined in connection with the contemporary phenomenon of re-familization, which entails people's increasing assumption of responsibility for taking care of their families and loved ones. At the same time, the broadening selection of communication devices and applications that extended families have at their disposal introduces new familial roles and responsibilities to ensure the proper functioning of the digital home. The claim is made that the concept of re-familization enables examination of these two phenomena in conjunction, especially in the Finnish context where the politics of re-familization has perhaps had more concrete bearings on individuals' lives than in many other places such as Italy and Slovenia, countries where public family benefits and services have never been as generously on offer or as extensively implemented.

© Springer Nature Switzerland AG 2019
S. Taipale, *Intergenerational Connections in Digital Families*,
https://doi.org/10.1007/978-3-030-11947-8_9

From De-familization to Re-familization

In the field of social policy, the notion of re-familization implies an about-face, a complete reversal of the politics of de-familization that once was the hallmark of the golden-era welfare state. Between the late 1950s and the mid-1980s, economic growth and low dependency ratios allowed expanding the public investments in family services and benefits in several European countries. This policy of de-familization was designed to promote adult citizens' ability to uphold a certain material standard and live independently of family support (Bambra, 2007; Esping-Andersen, 1999). To promote women's participation in the labour market, the expanding welfare states broadened the scope of their family services and benefits, especially in the fields of childcare and elderly care, and developed mechanisms to supply paid maternity leave (Daly, 2011).

The politics of de-familization had a particularly profound impact in Scandinavian societies, where the states favoured universal family (and other) benefits and services over more selective modes of welfare provision. In contrast, the Italian welfare system, for instance, traditionally always favoured family care networks over public services. Indeed, the country's care provision system has been left more or less unreformed even in the more recent times when the care needs of families have started piling up due to rising retirement age and the ageing of the population (Ranci & Sabatinelli, 2014), Slovenia, on the other hand, is often presented as a showcase example of a post-socialist country successfully transitioning to the market economy. For den Dulk et al. (2011), this has meant that in Slovenia, the re-consolidation of work and family has been regarded as a personal matter that shall be supported by the state rather than the employer and private organizations. In the country, the state's support for families was able to continue after the transition to capitalism in the 1990s, thanks to a comparatively good economic growth, successful social dialogue and a gradual transition process to the market economy. Despite the relatively successful transition period, however, the gap between the formulated policies and people's actual ability to claim the services and benefits remained wide. In the early 2000s, the combined effects of privatization, re-structuring and tightened international competition began to be felt in the Slovenian labour market and economy, undermining rights related to parenthood that had withstood since the socialist era (Kanjuo-Mrčela & Černigoj-Sadar, 2011).

Following a period of persisting austerity that brought with it significant cutbacks even in the wealthiest welfare states, the politics of re-familization were then introduced in Europe in the 2000s (Starke, 2006). In several countries on the continent, the eligibility criteria for claiming public family services like childcare or elderly home care assistance were tightened, and the scope of services was circumscribed. As the states pulled back, the concept of re-familization was introduced to describe the growing responsibility of families in organizing care and assistance for their members with particular needs (e.g. Kröger & Bagnato, 2017; Leira, 2002).

To understand how such care is, or could be, provided from afar and en route, scholars have for many years already focused their studies on families with small children and teenagers. Mobile communication technologies have proven invaluable and handy as tools for families to communicate both intra- and cross-generational intimacy within their sphere (Hjorth & Lim, 2012; Sawchuk & Crow, 2012). As has also been noted, specific rules concerning the use of technology (e.g. screen time restrictions) apply within the families, and unwritten rules concerning, for instance, the intra-family division of digital housekeeping chores, are not only tested and resisted but also contribute to family coherence and foster intimacy (Hjorth & Lim, 2012; Schofield Clark & Sywyj, 2012; Urry & Elliott, 2010). With the rise of the digital family, it has, moreover, turned out that the discussion, organization and monitoring of the daily help and care needs of older family members also take place using digital media and communication tools for the purpose (e.g. Petrovčič, Fortunati, Vehovar, Kavčič, & Dolničar, 2015; Tsai, Tsai, Wang, Chang, & Chu, 2010).

These sorts of family-initiated uses of digital media and communication technology that aim to resolve older family members' daily help and care needs are well in line with the aims of the European Union strategies and policy programmes viewing digital innovation as one way to empower the continent's citizenry in order to help keep older people healthy, independent and active (e.g. European Commission, 2017). Behind the catchword of citizen empowerment—a term hard for anyone to object to as such—there is, however, also an economic motive for promoting digitalization: the need to restrain public expenditure. Throughout Europe, digitalization of public services, including family, health and older-age care services, has been advanced parallel to governments' failure to provide publicly funded in-person help and care for families in need. The European Union has begun to promote more favourable conditions for lucrative e-health and telecare markets, stimulating also the growth of the so-called Silver Economy (e.g. European Commission, 2017, 2018; Ministry of Finance, 2018). In addition to the goodwill of the families, the EU seems to increasingly want to rely on markets as providers of technological innovations capable of making up for any shortcomings in, or otherwise shoring up, the public care provision in crisis. Combined with the vigorous promotion of e-health and remote-care technologies along with the digitalization of public services, it thus seems clear that any politics of re-familization can only presume even more solidarity in families than before and make family members even more dependent on one another's willingness to help in technological matters than what has been the case to date.

Technological Aspects of Re-familization[1]

The proliferation of mobile phones and personal computers in the 1980s and 1990s coincided with the spread and intensification of the politics of de-familization in

---

[1]Earlier versions of parts of this section were originally published in Hänninen, Taipale, & Korhonen (2018).

Europe and beyond. The rise of personal communication technology was predicted to lead to the dissolution of family solidarity, favouring as it was seen individual networking via person-to-person communication tools over more communal forms of interaction and communication (see, e.g. Rainie & Wellman, 2012). Mobile phones as the first genuinely mobile and portable communication tools brought with them the promise of the possibility to break loose from the binding ties of family. As Viken (2008) has pointed out, at the turn of the new millennium, it was even claimed that social networks more and more often rose as structures connecting specific roles, not persons as in traditional, densely knit families. Yet, perhaps the most striking example of how the politics of de-familization manifests itself in the practices of mobile communication technology use in families has to do with women's involvement in the labour market. Mobile phones were seen as particularly supportive of women's increased participation in the labour market, since they allowed them to manage their family affairs as well as their social and affectual relationships from a distance. The reverse side of this undeniable fact was, however, that the same mobile phone use also ended up reproducing many gender inequalities, as, in many places, it led to women's continuing to shoulder the main responsibility for family communication, even where men, too, armed with the exact same mobile communication tools, could have easily become more involved in the micro-coordination of family activities (see, e.g. Fortunati & Taipale, 2012; Rakow & Navarro, 1993).

The empirical materials discussed in this book are illustrative of the life of digital families more or less three decades after the introduction of the GSM standard and the commercialization of mobile phones. The evidence these materials offer suggests that in countries like Finland, where loosely connected extended families have been the norm for a long time already, new mobile and social media may have helped, at least somewhat, to revitalize family relationships. This change has been made possible by advances in personal media and communication technology that have opened up a possibility to engage in group-based communication using the equipment brought to the market. As we have seen, new one-to-many communication channels such as WhatsApp (see Chap. 7) open up entirely new ways of keeping a large number of family members, if not the entire family, connected (see also, e.g. Castells, 2010; Ling & Lai, 2016). Today, it no longer matters how large or geographically dispersed the family might be, as new group messaging and video conferencing technologies allow contacting all of its members at once, with no extra effort comparable to that required in one-to-one communication (Hänninen et al., 2018; Neustaedter, Harrison, & Sellen, 2013). Moreover, in technologically advanced countries, older family members' greater involvement in these new modes of communication has allowed their participation also in intra-family messaging groups and social media platforms. In this study, that was the case most prominently in the families in Finland. In contrast, especially in Slovenia, there was in many cases basically no need at all to engage in intensive online family interaction or chat group activities, as the three-generation families common in the country were socially tightly knit and physically closely connected anyway.

In the empirical analyses in this book, several ways for re-familization to manifest itself in the everyday life of digital families could be identified. First, members of digital families, especially in Finland, consistently found the new forms and channels of communication to have increased intra-family communication among them (see also Hänninen et al., 2018). The key informants who had moved away from home some years ago (such as the Finnish Emma, aged 24) pointed out how there had been considerably less communication among their family members earlier, before they had started using group-based communication tools and social media platforms. In particular, exchange of small messages among family members was seen as an act of caring for others. As family gatherings were not very frequent, it was thought of as important to know-how and what other family members were doing, wherever they happened to be or reside (e.g. the families of Maria, aged 24, and Marika, aged 20, in Finland). Moreover, when certain family members were not involved in the family's daily communication via digital technologies and applications, this was seen as something jeopardizing or directly undermining the unity of the family.

Second, given how the appropriation and maintenance of new digital technologies can impact the outlook and configuration of traditional family roles, re-familization can also be said to imply democratization of the family (see Hänninen et al., 2018). As older family members begin to rely, and even grow dependent, on younger members' expertise for technology purchases and, especially, assistance with software and application installation and maintenance, the family becomes functionally more consolidated and the voice of the young becomes better heard in it. Even though the economic authority of the family's breadwinner(s) was still emphasized even in this study, and the new responsibilities were not always experienced as unproblematic by the younger family members in it, all the informants, regardless of their age and generation, viewed this aspect of the re-familization as a development desirable for them.

Third, the rise of warm experts was another aspect of re-familization, one that was closely connected to the democratization of the family (see Hänninen et al., 2018). With ICTs becoming increasingly unavoidable as household items and essential for the smooth operation and effective management of the daily affairs of the family, warm experts had become an irreplaceable asset for many extended digital families. Indeed, even with the increasing intuitiveness and ease of use of the new products coming to the market, combined with the steady increase, across all age cohorts, in self-assessed digital skills over the years, the need for warm experts has not diminished in digital and other types of extended families (Olsson & Viscovi, 2018). Some explanations for this can be sought using a post-Mannheimian approach to generations that takes into account the intertwining of life-stage-specific needs and generation-specific ways of relating to new technologies (see Chap. 4). While it is, in general, the youngest and oldest people who are most dependent on the availability of external help, those in the oldest generations are typically also the ones most diverse as an age group, in terms of their physiological, psychological, social and functional traits (Nelson & Dannefer, 1992). What this 'aged heterogeneity' means is that, even if an entire generation would become digitally literate before it grows old, the help needs of many in it are nevertheless likely to increase steadily over time, leading to a greater variability in the help needs of older people as the unwanted effects of ageing, reflected also in one's ability to use new technologies, do not anyhow victimize all individuals equally.

Fourth, re-familization in this study manifested itself also in the use of time. Recognizing others' needs, and especially then taking care of them, requires time. As seen in Chap. 5, warm experts devoted considerable amounts of time to providing technical assistance, teaching digital skills and sorting out technical problems in their digital families. When families were geographically dispersed, apart from longer, in-depth phone and video calls, regular exchange of small messages was taken as a sign of caring and one's availability to others. Frequently exchanging short messages, family members could stay constantly connected and maintain their sense of togetherness (cf. Cao, 2013). Family messaging thus provided a good example of the ways in which digital families could be actively 'done' through mobile communication.

Fifth, re-familization, to a certain extent at least, also meant increased internal solidarity for the digital families. In them, solidarity, understood as a strong sense of personal duty towards others (Ter Meuler & Wright, 2012), was manifested especially in a sense of responsibility for ensuring the proper functioning of new technologies and solving other family members' technical problems. This aspect of re-familization was particularly pronounced in Slovenia, where family ties were close and where the informants more often than elsewhere suggested that family members had a duty to help one another in the use of new technology. Despite clear country-specific differences on this issue, however, the feelings of solidarity were mainly related to the functional and associational aspects of digital technology use. This is in line with Peng et al. (2018) notion of digital solidarity, a term coined as an extension to association and functional solidarity when analysing mothers' attempts to stay connected with their grown-up children.

Sixth, it is worth reminding that re-familization is not always a positive process affecting everyone fairly or similarly. As Hänninen et al. (2018) have pointed out based on the same research material as that examined here, communication in digital families tends sometimes to become compartmentalized. In such cases, only family members with the necessary devices, right applications and sufficient digital skills, or those sharing the same communication style and preferences, get connected with one another digitally. Most often, the compartmentalization within the distributed extended families in this study meant that fathers and/or grandparents were left outside the circle of younger family members and their mother. These 'excluded' family members could nevertheless be active digital communication technology users, and hence otherwise be part of the digital family; it was only their drastically different or very limited communication practices that kept them in the outer circle of family's communication community. Another example of the uneven effects of familization was the asymmetrical distribution of the costs and benefits of help provision. The role and tasks of the warm expert tended to fall upon just one or two members of the family, making all others in it basically pure beneficiaries.

Seventh, the way and extent to which the above six aspects of re-familization were visible in the three countries in this study varied considerably. As already seen, this variation was due to the prevailing family structure and housing arrangements, the level of intergenerational solidarity, as well as families' preparedness to use different media and communication technologies in each case. Although the families in all of these countries had more or less the same range of digital technologies and applications available to their use, they used the different affordances of these devices and applications accommodating them to their own country context. In Finland, where even the smallest families were geographically extremely scattered, families benefited most from group messaging and social media platforms that helped them reinforce and even revitalize family ties. In Italy, on the other hand, it was larger family networks involving cousins, aunts, uncles and even overseas family members that provided the stimulus for adopting new communication technologies. Even though the key informants in Italy did not as frequently as those in Finland live independently of their parents, they had larger circles of family members with whom to stay connected. Finally, Slovenia turned out to be a special case in many respects. In this small country, families were typically geographically concentrated, with all family members living in the same narrowly circumscribed area or even the same building. As a result of this family members' close proximity to one another, the Slovenian participants in this study expressed fewer needs for technology use that could foster family coherence or family unity. Frequent daily encounters with other family members, particularly older relatives living in the same building or on the same property, lent themselves especially well for fluent and everyday intergenerational counselling and instruction on digital technology use and maintenance. With these and other country specificities outlined in this book in mind, it would then be misleading and inaccurate to presume re-familization to have unfolded in, and affected, all the three countries, and beyond, to the same degree and in the same fashion.

References

Bambra, C. (2007). Defamilisation and welfare state regimes: A cluster analysis. *International Journal of Social Welfare, 16*(4), 326–338.

Cao, X. (2013). Connecting families across time zones. In C. Neustaedter, S. Harrison, & A. Sellen (Eds.), *Connecting families: The impact of new communication* (pp. 127–139). London: Springer.

Castells, M. (2010). *Rise of the network society. The information age: Economy, society, and culture I* (2nd ed.). Malden: Wiley-Blackwell.

Daly, M. (2011). What adult worker model? A critical look at recent social policy reform in Europe from a gender and family perspective. *Social Politics: International Studies in Gender, State & Society, 18*(1), 1–23.

den Dulk, L., Peper, B., Černigoj-Sadar, N., Lewis, S., Smithson, J., & Van Doorne-Huiskes, A. (2011). Work, family, and managerial attitudes and practices in the European workplace: Comparing Dutch, British, and Slovenian financial sector managers. *Social Politics, 18*(2), 300–329.

Esping-Andersen, G. (1999). *Social foundations of postindustrial economies*. Oxford: Oxford University Press.

European Commission, EC. (2017). *Growing the Silver Economy in Europe*. Retrieved from https://ec.europa.eu/digital-single-market/en/news/growing-silver-economy-europe.

European Commission, EC. (2018). *Digital single market*. European Commission. Retrieved from https://ec.europa.eu/digital-single-market/.

Fortunati, L., & Taipale, S. (2012). Women's emotions towards the mobile phone. *Feminist Media Studies, 12*(4), 538–549.

Hänninen, R., Taipale, S., & Korhonen, A. (2018). Re-familization in the broadband society. The effects of ICTs on family solidarity in Finland. *Journal of Family Studies*. Advance online publication. https://doi.org/10.1080/13229400.2018.1515101.

Hjorth, L., & Lim, S. S. (2012). Mobile intimacy in an age of affective mobile media. *Feminist Media Studies, 12*(4), 477–484.

Kanjuo-Mrčela, A., & Černigoj-Sadar, N. (2011). Social policies related to parenthood and capabilities of Slovenian parents. *Social Politics, 18*(2), 199–231.

Kröger, T., & Bagnato, A. (2017). Care for older people in early twenty-first century Europe. In F. Martinelli, A. Anttonen, & M. Mätzke (Eds.), *Social services disrupted* (pp. 201–218). Cheltenham: Edward Elgar.

Leira, A. (2002). *Working parents and the welfare state: Family change and policy reform in Scandinavia*. Cambridge: Cambridge University Press.

Ling, R., & Lai, C. H. (2016). Microcoordination 2.0: Social coordination in the age of smartphones and messaging apps. *Journal of Communication, 66*(5), 834–856.

Ministry of Finance. (2018). *Digitalisation*. Retrieved from https://vm.fi/en/digitalisation.

Nelson, E. A., & Dannefer, D. (1992). Aged heterogeneity: Fact or fiction? The fate of diversity in gerontological research. *The Gerontologist, 32*(1), 17–23.

Neustaedter, C., Harrison, T., & Sellen, A. (Eds.). (2013). *Connecting families: The impact of new communication technologies on domestic life*. Dordrecht: Springer.

Olsson, T. & Viscovi, D. (2018, November) Warm experts for elderly users: Who are they and what do they do? *Human Technology, 14*(3), 324–342.

Peng, S., Silverstein, M., Suitor, J. J., Gilligan, M., Hwang, W., Nam, S., et al. (2018). Use of communication technology to maintain intergenerational contact: Toward an understanding of 'digital solidarity'. In B. B. Neves & C. Casimiro (Eds.), *Connecting families? Communication Technologies, generations, and the life course* (pp. 159–180). Bristol: Polity.

Petrovčič, A., Fortunati, L., Vehovar, V., Kavčič, M., & Dolničar, V. (2015). Mobile phone communication in social support networks of older adults in Slovenia. *Telematics and Informatics, 32*(4), 642–655.

Rainie, L., & Wellman, B. (2012). *Networked: the new social operating system*. Cambridge: MIT Press.

Rakow, L. F., & Navarro, V. (1993). Remote mothering and the parallel shift: Women meet the cellular telephone. *Critical Studies in Media Communication, 10*(2), 144–157.

Ranci, C., & Sabatinelli, S. (2014). Long-term and child care policies in Italy between familism and privatisation. In M. Leon (Ed.), *The transformation of care in European societies* (pp. 233–255). London: Palgrave Macmillan.

Sawchuk, K., & Crow, B. (2012). "I'm G-Mom on the Phone": Remote grandmothering, cell phones and inter-generational dis/connections. *Feminist Media Studies, 12*(4), 496–505.

Schofield Clark, L., & Sywyj, L. (2012). Mobile intimacies in the USA among refugee and recent immigrant teens and their parents. *Feminist Media Studies, 12*(4), 485–495.

Starke, P. (2006). The politics of welfare state retrenchment: A literature review. *Social Policy & Administration, 40*(1), 104–120.

Ter Meulen, R., & Wright, K. (2012). Family solidarity and informal care: The case of care for people with dementia. *Bioethics, 26*(7), 361–368.

Tsai, H. H., Tsai, Y. F., Wang, H. H., Chang, Y. C., & Chu, H. H. (2010). Videoconference program enhances social support, loneliness, and depressive status of elderly nursing home residents. *Aging and Mental Health, 14*(8), 947–954.

Urry, J., & Elliott, A. (2010). *Mobile lives*. London and New York: Routledge.

Viken, A. (2008). The Svalbard transit scene. In J. O. Barenholdt & B. Granas (Eds.), *Mobility and place: Enacting northern peripheries* (pp. 139–154). Aldershot: Ashgate.

Chapter 10
Towards a More Balanced Approach
to Digital Families

Abstract The book concludes with the claim that the modes and frequency of intra-family digital communication cannot be studied separately from the social functions that the different technologies have in extended families. In digital communication, when problems related to the use of new technology arise, a caring relationship emerges between a carer, attentive to the expressed care needs of the cared-for, and the latter, expected to provide some response in exchange for the help received. Finally, avenues for future research to are outlined, with the future of the digital family briefly considered.

Keywords Caring relationship · Digital family · Digital technology · Warm expert

This book has explored intergenerational connections in digital families from various angles, comparing the situation in families living in Finland, Italy and Slovenia. As was found, various improvements in digital media and communication technologies, the spread of digital skills across generations, the digitalization of the home, and, especially, the more extensive employment of mobile communication technology and social media for the purposes of intra-family communication have had the joint outcome of allowing families to experience more connectivity, more togetherness and more unity across generational boundaries than before. In the digitally connected families participating in this study, this was largely experienced as a positive, although not entirely unproblematic, development.

Previous research highlighting the positive impacts of ICT on family relationships has mainly revolved around transnational family relationships, in which context both conventional ICTs (such as voice telephony and text messaging) and newer forms of digital media (social networking sites) have been presented as a lifeline helping people to stay connected (Pham & Lim, 2016). As studies have found, shared deployment of digital communication tools enables maintenance of a sense of presence in diasporic families (Baldassar, 2008; Yoon, 2016), allows parenting from afar (Chib et al., 2014; Madianou & Miller, 2011, 2012) and enacts what has been called 'friendly surveillance', or the performance of care rather than monitoring activities, within the family (Sinanan & Hjorth, 2018). The overall positive view taken in these studies of the role of ICTs stems from the fact that in transnational families, digital technologies are (rightly so) not regarded as a cause for the dispersion of the family;

© Springer Nature Switzerland AG 2019
S. Taipale, *Intergenerational Connections in Digital Families*,
https://doi.org/10.1007/978-3-030-11947-8_10

quite the contrary, they are seen as vehicles for compensating for the high price paid by the families for being physically separated.

More typically, however, research has presented digital technologies as something akin to a double-edged sword. On the one hand, especially personal media and personal communication tools have been seen as enablers of more individualized and mobile lifestyles, allowing more personalized daily agendas and schedules. From this point of view, family appears as an increasingly loose network of more and more individually networked members (e.g. Kennedy & Wellman, 2007). On the other hand, though, mobile and other personal communication tools have been shown to help the new, networked families to micro-coordinate and manage their daily activities on a constant basis (e.g. Ling & Yttri, 2002; Neustaedter et al., 2013). In other words, the technologies have been seen as both eroding the social coherence of families once fostered by physical proximity and the bounds of locality, and providing new means for managing family relationships irrespectively of time and space.

At the same time, however, there is also a large body of research presenting a more critical view according to which the growing dominance of new digital media, combined with the increasingly more individualized nature of communication, drains social relationships of emotions and intimacy. It has been proposed, for instance, that communication in virtual environments produces an illusion of companionship and trustful relationships with no presumption of any emotional or longer term commitment (e.g. Turkle, 2011). Especially, earlier studies of Internet use typically ended up lamenting the diminishing time spent together in families under the impact of digital technologies and media contents consumed in isolation (e.g. Nie & Erbring, 2002). Newer studies, however, have found very little or no support at all for this particular argument. Vriens and van Ingen (2018), for instance, were able to conclude that the decreasing number of strong social ties and the quality of online relationships were a far more serious concern than the time spent in interacting with one's close relationships. Along the same lines, Vilhelmson, Thulin, and Elldér (2017), examining the results of a Swedish time use survey from 2010 to 2011, showed that the time spent on ICT use was not directly away from the interaction with family members.

Compared to previous research, this book has attempted a more balanced view on the use of media and digital technologies in families. It has shown families to greatly appreciate their improved possibilities to keep in touch with more family members, along with the hard work carried out by the warm experts for the common good of all family members. In all three counties, Finland, Italy and Slovenia, extended families differently but firmly bonded together through, and in close connection with, digital media and communication technologies. While in Finland families used instant messaging applications and social media platforms to breathe life into their intra-family communication, in Italy families were still on the verge of adopting a more diverse set of digital communication tools for their intra-family interaction. In Slovenia, for contrast, families were tied together through notably close helping relationships, operative in technology use as well and made possible by the physical proximity of others in one's shared everyday life.

As the findings above further indicate, digital families in all the three countries had accepted it as a fact that digital media and communication technologies had permanently made their way into family life. Regardless of whether the family bonding and unity were the result of intensified technology-mediated intra-family communication or could be attributed to engagement in cross-generational help provision in technology use, there were no signs of any strong resistance towards the digitalization of family life in any of the families considered, in any of the countries involved. This was so even when older family members' understanding of what constituted the 'proper' ways of using digital technology deviated from the views and perceptions held by the younger ones, which indeed was the case almost every time (cf. Colombo, Aroldi, & Carlo, 2018). Even such differences were not reported as having any significance as possible sources of family conflicts or disagreements. Even in families where there were members purposefully using only a smaller array of new technologies, or who deliberately limited the time they spent on technology use, no one was reported to yearn for the good old days, before digital technology arrived. In general, family members' in-depth knowledge of one another's desire and also actual tendency to adjust their technology and technology use to that of everyone else's was one contributing factor in the development of a sense of unity across family generations.

From Connections to Caring Relationships

What the findings and the discussion in this book suggest is that it no longer suffices to study family digital connections, or, the modes and frequency of intra-family digital communication, in isolation of the social functions that the different technologies have in extended families. The question should, however, also be asked as to why digital families invest so much effort in trying to have family members be able to reach one another to maintain and reinforce family relationships, as we have seen them do in this study. Why do family members help one another, sometimes notably altruistically, to acquire, take into use, and actually deploy new digital technologies, even when doing so may be experienced as not just tiresome, but also demanding and difficult, with no immediate benefit to oneself or guarantees of any long-term learning outcomes?

As soon as the focus of the enquiry is moved away from the density and frequency of intra-family connections to the quality and social functions of the intergenerational communication, it becomes obvious that the digital family is much more than the sum of its digitally connected individual members. From the latter viewpoint, it appears that the use of digital technologies in families is to a great extent about maintaining familial caring relationships, both across and within generations. Although to a large extent a still-unexplored territory, the question of the various uses of digital media and communication technologies is clearly entwined with the issue of intimacy and caring in family life (Baldassar, 2016; Sinanan & Hjorth, 2018).

In general, caring relationships serve people's daily life, helping them to meet their daily needs ranging from material and bodily to mental and social ones (cf. Fischer & Tronto, 1991). Accordingly, having a caring relationship entails listening to the other person's needs, engaging in a dialogue, critical thinking and reflection, and showing responsiveness (Noddings, 2012). As we have seen, in digital communication and with problems related to the use of new technology, a caring relationship emerges between a carer (often a warm expert), who is attentive to the (sometimes rather implicitly) expressed care needs of the cared-for, and a cared-for, who is expected to provide some response in exchange for the help received. As was also suggested above, the two parties may also switch positions in the course of time and reverse their caring relationship. Although the role of the carer is often assigned to the young warm expert(s) in the family, adults and grandparents care for their children as well, for instance, by teaching them how to get started with their first digital devices and services and by looking after them and monitoring their online behaviour.

As this study found, in countries like Finland, where families are highly dispersed and individualized, caring relationships are increasingly played out and experienced through digital communication technologies. In families where this is so, opportunities for physically coming into direct contact with other family members' needs are more limited and infrequent. It also worth noting that the actual information content of intra-family online communications is often of secondary value only: the most important need that these communications serve is simply to know that others in the family are doing well. The exchange of seemingly unimportant messages (see Chap. 7), the making of short, trivial telephone calls and the liking of other family members' social media posts all serve to sustain caring relationships. Moreover, as this book has also shown, such regular and frequent digital connections with other family members are not the only, and sometimes not even the most important, way of expressing caring in the digital family. Taking care of others in it also manifests itself as readiness to provide hands-on help when others encounter problems with digital devices, applications or online services.

In a wider societal context, we might make the observation that caring relationships in digital families resonate with, and take shape in response to, the politics of re-familization discussed above (Chap. 9). Following Tronto's (1994) four phases of care provision, it could be argued that, for people living in such, the digital family provides a primary context for help provision in technological matters. There are several reasons for why this should be so. To begin with, an extended family provides a natural environment for *caring about*. Caring about refers to being attentive to the needs of others, whether the question is of basic needs such as for food and safety or, as in a more modern-day technological context, higher order needs arising from the use of ordinary digital technologies. Second, as we have already seen above, it is typically the warm experts in the family who *take care* of others, meaning that they often feel personally responsible for the proper functioning of the digital technologies in the possession of their family. Moreover, warm experts are typically in charge of *caregiving*, which, in connection with technology use, is about provision of technical assistance in problem-solving. In that role, warm experts serve to ultimately fulfil the digital needs of others. Lastly, caring relationships also contain the element of *care receiving*. As also observed above, warm experts, as the digital caregivers in the family, are very sensitive to the reactions of their help receivers. Correspondingly, also the help receivers spend time and energy in thinking how their requests for help might be perceived and received by the warm experts whom they approach.

There are many avenues for further research to investigate how caring relationships are played out in practice in the digital family. First of all, there is a need to clarify what, in extended digital families, facilitates the recognition of others' needs in technology use. We appear to know already now, however, that the members of digital families are, on the whole, relatively well aware of the other family members' preferred modes of contacting one another and of their ability to employ different types of communication devices and applications. In this particular regard, locally and nationally distributed digital families have an advantage over transnational families characterized by a more permanent physical separation of their members: the better opportunities they offer for in-person encounters and family reunions allow for closer monitoring of the development of family members' technical skills and communicative preferences. While remote provision of technical assistance, such as by telephone or via video link, is often considered awkward, regular or even occasional visits instead make it possible to request and provide hands-on help in technological matters in person, and without prior consultation or major arrangements.

Second, to date only very little has been studied regarding the responsibilities felt for helping other family members in technology use. In this book, we saw that warm experts, the persons considered also by other family members as responsible for the proper functioning of digital technology in the family, are quite expressly singled out in families. Nevertheless, warm experts themselves experience their responsibility not solely as a burden but also as something rewarding to them. It is, moreover, also worth keeping in mind that the responsibilities of the warm experts are not fixed but subject to change and redistribution as families age. Furthermore, with older family members busy becoming digitally more versed, yet not any more immune to the physiological and cognitive effects of ageing than before, we also need to learn more about how, and to what extent, warm experts' responsibilities are passed on from one generation to the next as digital families grow older.

The third question that has largely fallen under researchers' radar concerns any possible positive long-term effects of the help and care provided by warm experts. In some studies, the use of smartphones for caring for others was associated with lower levels of loneliness and depression and higher levels of self-esteem in the caregivers (e.g. Park & Lee, 2012). It might therefore be that digital families benefit from more intensive, intimate and caring family relationships in more diverse and nuanced ways than what research has so far been able to find out. At any rate, what remains obvious is that any positive outcomes in helping relationships are, in general, only possible insofar as both of the parties to that relationship, that is, both the carer and the cared-for, experience the relationship as mutually beneficial and rewarding.

The Future of the Digital Family

The digital family is already reality in some countries like Finland that led the way when mobile and personal communication technologies were first introduced and began to be appropriated on a large scale. In countries that are latecomers to digitalization, it will still take some years before the oldest members of extended families will be able to embrace smartphones and mobile Internet connectivity in such a large scale as to enable their entire large families to benefit from everyone's being online and digitally connected. As already noted, moreover, it also needs to be kept in mind that extended families in Europe (and beyond) are transforming into the digital families at a different pace and following different paths. While a certain particular technology or application may serve the everyday needs of families in one country, it may be experienced as impractical or entirely inappropriate in another. What remains uncertain for the time being is, furthermore, whether there will be some leapfrogging technologies or applications that can help digitally less equipped and less versed families catch up with those ahead of them in the developments.

Regarding the future of the digital family, it is the oldest members of families who are in a crucial position. First of all, although older people are often considered as reluctant technology adopters, there are many in that group who are even now busy taking up new digital technologies. Research on the subject should, accordingly, be prepared to acknowledge that any connection between age and technology adoption or use is rather nonlinear than linear. The kind of post-Mannheimian approach to technology user generations as outlined in this book (Chap. 4) offers one theoretical framework for doing so and using the insight for the benefit of future research. Second, as previous research has shown, older technology users are a highly diverse group, in terms of individuals' functional characteristics, skills levels, personality traits, personal history of technology use and support networks available—indeed, more notably so than the younger groups of users (see, e.g. Sourbati, 2015). With the almost inevitable increase in physical and cognitive impairments with age, however, almost everyone in that group nevertheless undergoes changes in their functional abilities that impact their facility in using digital technologies and applications, given that the latter typically require good vision and hearing as well as steady hands and fine motoric skills for their operation. Indeed, studies have already shown poor health condition to be a stronger determinant for older people's low engagement with intra-family communication than their generational membership per se (e.g. Peng et al., 2018). Third, as evident also from this book, almost all of digitally mediated family communication, whether dyadic or group-based, occurs between two consecutive generations. Thus, it would seem to be of pertinence to study factors promoting the kind of skipped-generation communication noted earlier, or, interaction that brings together children and grandparents without parents' involvement. The absence of such skipped-generation communication does not, however, indicate a complete lack of direct interaction between children and grandparents. Yet, as the results of this study suggest, any such interaction is likely to be about face-to-face type of assistance given in technology use contexts.

To conclude, the future of the digital family is not simply shaped by technologic advancements and innovations alone. It is much more dependent on the ways in which increasingly varied families appropriate and make use of the rapidly changing landscape of digital devices, programmes and applications. Depending on cultural expectations and the prevailing social norms, digital families may either end up reproducing existing social inequalities in family life, such as those based on gendered practices of family communication or unequal division of household chores, or, at the best, promoting a more democratic and inclusive family culture through new technologies that are supportive of re-familization.

References

Baldassar, L. (2008). Missing kin and longing to be together: Emotions and the construction of co-presence in transnational relationships. *Journal of Intercultural Studies, 29*(3), 247–266.

Baldassar, L. (2016). Mobilities and communication technologies: Transforming care in family life. *Family life in an age of migration and mobility* (pp. 19–42). London: Palgrave Macmillan.

Chib, A., Malik, S., Aricat, R. G., & Kadir, S. Z. (2014). Migrant mothering and mobile phones: Negotiations of transnational identity. *Mobile Media & Communication, 2*(1), 73–93.

Colombo, F., Aroldi, P., & Carlo, S. (2018). "I use it correctly!": The use of ICTs among Italian grandmothers in a generational perspective. *Human Technology, 14*(3), 343–365.

Fischer, B., & Tronto, J. (1991). Towards a feminist theory of care. In E. Abel & M. Nelson (Eds.), *Circles of care: Work and identity in women's lives* (pp. 35–54). Albany: State University of New York Press.

Kennedy, T., & Wellman, B. (2007). The networked household. *Information, Communication & Society, 10*(5), 645–670.

Ling, R., & Yttri, B. (2002). Hyper-coordination via mobile phones in Norway. In K. Katz & M. Aakhus (Eds.), *Perpetual contact: Mobile communication, private talk, public performance* (pp. 139–169). Cambridge: Cambridge University Press.

Madianou, M., & Miller, D. (2011). Mobile phone parenting: Reconfiguring relationships between Filipina migrant mothers and their left-behind children. *New Media & Society, 13*(3), 457–470.

Madianou, M., & Miller, D. (2012). *Migration and new media: Transnational families and poly-media*. London & New York, NY: Routledge.

Neustaedter, C., Harrison, T., & Sellen, A. (Eds.) (2013). *Connecting families: The impact of new communication technologies on domestic life*. Dordrecht: Springer

Nie, N. H., & Erbring, L. (2002). Internet and society: A preliminary report. *IT & Society, 1*(1), 275–283.

Noddings, N. (2012). The caring relation in teaching. *Oxford Review of Education, 38*(6), 771–781.

Park, N., & Lee, H. (2012). Social implications of smartphone use: Korean college students' smartphone use and psychological well-being. *Cyberpsychology, Behavior, and Social Networking, 15*(9), 491–497.

Peng, S., Silverstein, M., Suitor, J. J., Gilligan, M., Hwang, W., Nam, S., et al. (2018). Use of communication technology to maintain intergenerational contact: Toward an understanding of 'digital solidarity'. In B. B. Neves & C. Casimiro (Eds.), *Connecting families? Communication technologies, generations, and the life course* (pp. 159–180). Bristol: Polity.

Pham, B., & Lim, S. S. (2016). Empowering interactions, sustaining ties: Vietnamese migrant students' communication with left-behind families and friends. In Lim, S. S. (Ed.), *Mobile Communication and the Family* (pp. 109–126). Dordrecht: Springer.

Sinanan, J., & Hjorth, L. (2018). Careful families and care as 'kinwork': An intergenerational study of families and digital media use in Melbourne, Australia. In B. B. Neves & C. Casimiro (Eds.), *Connecting families? Communication technologies, generations, and the life course* (pp. 181–200). Bristol: Polity.

Sourbati, M. (2015). Age(ism) in digital information provision: The case of online public services for older adults. In J. Zhou & G. Salvendy (Eds.), *Human aspects of IT for the aged population* (pp. 376–386). Dordrecht: Springer.

Tronto, J. (1994). *Moral boundaries: A political argument for an ethic of care*. New York and London: Routledge.

Turkle, S. (2011). *Alone together: Why we expect more from technology and less from each other*. New York: Basic Books.

Vilhelmson, B., Thulin, E., & Elldér, E. (2017). Where does time spent on the Internet come from? Tracing the influence of information and communications technology use on daily activities. *Information, Communication & Society, 20*(2), 250–263.

Vriens, E., & van Ingen, E. (2018). Does the rise of the Internet bring erosion of strong ties? Analyses of social media use and changes in core discussion networks. *New Media & Society, 20*(7), 2432–2449.

Yoon, K. (2016). The cultural appropriation of smartphones in Korean Transational Families. In Lim, S. S. (Ed.), *Mobile Communication and the Family* (pp. 93–108). Dordrecht: Springer.

Appendix

A considerable part of this book is based on materials collected between 2014 and 2015 in an Academy of Finland-funded project entitled 'Intergenerational Relations in Broadband Societies', carried out in Finland, Italy and Slovenia. The method used for the data collection was, subsequently, termed as Extended Group Interviews (EGI; see Hänninen, Taipale, & Korhonen, 2018). This interview technique enabled the study of entire extended families, notwithstanding the fact many of the families involved in this study were geographically dispersed and consisted of multiple households. This appendix describes the study participants, the data collection method and the analytical techniques used.

Informants and Key Informants

The main material for this book consisted of 66 written reports collected from informants in the three countries studied in 2014 and 2015. College students from three universities served as *key informants*. They observed and interviewed their own family members, who acted as *informants* representing three and even four family generations.

The key informants were recruited from three different universities, one in each country. A key criterion in their selection was that they all have a similar academic background relevant to this research. That is to say, they were all to either have taken or be currently enrolled in a study programme or a separate course capable of providing them with a sufficient basic knowledge of social-scientific research on new media and communication technologies. This method of using college students as key informants had several practical advantages. All of them, for instance, were familiar with basic interviewing methods and had extensive experience in writing research papers and reports such as case-study essays and learning diaries and using such as part of their studies.

© Springer Nature Switzerland AG 2019
S. Taipale, *Intergenerational Connections in Digital Families*,
https://doi.org/10.1007/978-3-030-11947-8

In Finland, the key informants were social science and communications studies students from the University of Jyväskylä, recruited through university emailing lists. In Italy, undergraduate and graduate students enrolled in the Multimedia Communication study programme at the University of Udine were invited to take part in the study. As in Finland, this invitation was extended via an email list. In Slovenia, the key informants were students from the graduate programme in Social Informatics at the University of Ljubljana. The Slovenian key informants conducted their interviews and wrote their key informant reports as part of their coursework for a course they were enrolled in at the time of the study.

The key socio-demographic characteristics of the key informants and the informants that they interviewed and observed are presented in Table A.1. There were a total of 331 informants and 66 key informants in the three countries studied. The number of the key informants and their family informants was approximately the same in each country. In terms of their gender distribution, about two-thirds of the key informants were women (45), with, overall, a better balance in this regard in Italy than in Finland and Slovenia. The key informants in Finland and Slovenia were on average 4 years older than those in Italy (28–24 years, respectively). All in all, the key informants interviewed and observed a total of 162 female and 168 male family members. The gender balance among them, in other words, was quite good.

Table A.1 Characteristics of key informants and informants, by country

	Finland	Italy	Slovenia
Key informants (n)	22	21	23
Gender (n)			
Male	3	10	8
Female	19	11	15
Age (years)			
Range	20–38	21–28	23–30
Mean	28	24	28
Informants (n)	111	104	115
Gender (n)			
Male	50	51	61
Female	61	53	54
Relationship with the key informant (n (mean distance))			
Parent	36 (150 km)	31 (111 km)	42 (58 km)
Sibling/stepsibling	26 (217 km)	18 (219 km)	27 (141 km)
Grandparent	10 (239 km)	21 (164 km)	22 (90 km)
Other	39	34	25

Source Hänninen, Taipale, and Korhonen (2018)

The key informants were also asked to provide information on whether or not they shared the same household with their family informants. The average distances separating the key informants from their elsewhere-living interviewees were the shortest in Slovenia and the longest in Finland, where the key informants most often lived in households of their own. In Italy, the distances were very similar to those in Finland. In the Italian case, however, these figures were heavily tilted by a group of key informants (amounting to one-fifth of them) who lived very far (800–2000 km) away from the rest of their family. Compared to their Italian and Slovenian counterparts, the Finnish key informants lived especially far from not only their grandparents but also parents. The distance between the key informants and their siblings was long in all three countries, ranging from 141 to 219 km on average.

Research Procedure

The Extended Group Interview (EGI) method was designed to investigate inter-generational relationships among a relatively large number of family members. EGI is anchored into the tradition of collaborative ethnographic enquiry and new methodological openings in the field of family group interviews (Reczek, 2014). The attribute 'extended' refers to several special features of the EGI method. First, it underlines the fact that the method enables the study of extended multi-household families, instead of just single-household nuclear families. Second, it points to various methods of conducting interviews accommodated by EGI, ranging from in-person to technology-mediated interviews via phone, Skype and so forth. Third, EGI allows reaching a large number of family members, by extending the interviews from one specific time and place into a whole series of interviews (Hänninen et al., 2018).

EGI is, further, characterized by a collaborative element between the key informants and the main researcher. The key informants act as co-researchers who not only collect interview data for the main researcher(s) but also create their own interpretations of the data gathered when reporting back to the main researcher(s). This collaborative element in EGI allows, and even encourages, interviewees to freely express their own views (cf. Lassiter & Campbell, 2010; Rappaport, 2008). As one sign of the fact that this element indeed worked in the study as intended, key informants in it frequently included their dissenting voices in the reports they submitted to the main researcher.

Nevertheless, also key informants' pre-conceptions and prejudices can be assumed to have influenced their observations and interviews, at least to a certain extent (Marshall, 1996). There were also some other limitations inherent in the procedure. The key informants' double role as both a researcher and an informant, for instance, may have complicated their interactions with their family members. In addition, the key informants and their family members in the sample were ethni-cally rather homogeneous as a group. Including, for instance, ethnic minorities and immigrants in the sample would very likely have yielded information about practices of digital technology use in families that now remained undisclosed.

In each country, the key informants were given the exact same assignment: to observe ICT-related communication in their families for a period of 1 week, and then interview at least five of their family members on their use of digital media and communication technologies. Following this fieldwork period of theirs, the key informants wrote three reports with minimum of 300 words each, in which they were asked to answer the following questions: (1) Which ICT tools and applications were used in their families to stay in touch with other family members? (2) How would they assess their family members' relative ICT skills? (3) How had ICT shaped the different roles their family members had in the family? For the key informants, ICT was here defined, quite broadly, as all the different kinds of digital communication devices and services used to stay in contact and communicate with family members (including, e.g. mobile phones, email, Facebook, Twitter, WhatsApp and Instagram).

The key informants were instructed to interview at least one of their parents and one grandparent, if possible. They were free to determine the three remaining interviewees, provided that these would be of different ages. Some key informants thus interviewed their cousins, their children and their spouse's relatives. The key informants also compiled background information on the interviewees (their gender, age, relationship to the key informant, their geographical distance from the key informant if they did not share the same household). Also, the methods of data collection the key informants used for their different informants were reported.

In preparation for the EGI interviews, instruction sessions were organized in each university to inform their students about the research and its aims, and to obtain the informed consent of those of them recruited as key informants. The key informants were also informed about their right to withdraw from the study at any time without consequences. Each key informant had a contact person in her or his country who was available for questions and advice at all stages of the study. They were also informed of the fact that, to protect both their own and their family members' privacy, all names in published and unpublished work resulting from the study would be changed to pseudonyms. The key informants received a one-time honorarium of EUR 50 upon completing their assignments.

Analytical Tools

The research material was examined and analysed using two standard analytical techniques for qualitative interview data. First, for chapters built around an established theoretical framework (Chap. 8, on Bengtson and Roberts's model), certain concepts (Chap. 7, on 'reach' and 'phatic communion') or new conceptual categorizations (Chap. 6, on 'digital housekeeping'), the principles of a directed approach to qualitative content analysis were followed (see Hsieh & Shannon, 2005). This method takes a theory or incomplete findings from previous research as its starting point to help guide the initial coding. The initial categories obtained in this study were then re-examined to promote clustering around common themes. The strength of this approach is that it can provide support for the existing theories and concepts while at the same time helping to identify their shortcomings and limitations.

Next, when there were no clear pre-conceptions or strong theories guiding the analytical work, the research material was examined using thematic analysis of coded research data (see Boyatzis, 1998; Braun & Clarke, 2006). This method was applied, in particular, when the role of warm experts in digital families was explored (Chap. 5). In the first phase, the analysis of the reports was focused on the changes in family roles likely associated with new media and communication technologies. After that, the reports were analysed again focusing on possible connections between the life course and the daily chores of the warm experts. Third, the analysis turned the question of who in the families served as their warm experts and for whom. Lastly, the research material was reviewed once more, this time focusing on possible factors that motived warm experts in their work, and on the question of what made their work difficult/easy or taxing/rewarding.

References

Boyatzis, R. E. (1998). *Transforming qualitative information: Thematic analysis and code development*. London: Sage.

Braun, V., & Clarke, V. (2006). Using thematic analysis in psychology. *Qualitative Research in Psychology, 3*(2), 77–101.

Hänninen, R., Taipale, S., & Korhonen, A. (2018). Refamilisation in the broadband society. The effects of ICTs on family solidarity in Finland. *Journal of Family Studies*. Advance online publication. https://doi.org/10.1080/13229400.2018.1515101.

Hsieh, H. F., & Shannon, S. E. (2005). Three approaches to qualitative content analysis. *Qualitative health research, 15*(9), pp. 1277–1288.

Lassiter, L. E., & Campbell, E. (2010). What will we have ethnography do? *Qualitative Inquiry, 16*(9), 757–767.

Marshall, M. N. (1996). The key informant technique. *Family Practice, 13*(1), 92–97.

Rappaport, J. (2008). Beyond participant observation: Collaborative ethnography as theoretical innovation. *Collaborative Anthropologies, 1*, 1–31.

Printed in the United States
By Bookmasters